The Essence of
Artificial Intelligence

The Essence of Artificial Intelligence

Edited by
Xavier Smith

www.willfordpress.com

Published by Willford Press,
118-35 Queens Blvd., Suite 400,
Forest Hills, NY 11375, USA

ISBN: 978-1-68285-481-5

Cataloging-in-Publication Data

The essence of artificial intelligence / edited by Xavier Smith.
p. cm.
Includes bibliographical references and index.
ISBN 978-1-68285-481-5
1. Artificial intelligence. 2. Fifth generation computers. 3. Neural computers.
I. Smith, Xavier.
TA347.A78 E87 2018
006.3--dc23

For information on all Willford Press publications
visit our website at www.willfordpress.com

Contents

Preface

Artificial intelligence is one of the most important sub-fields of computer science in the present scenario. It refers to the study of intelligence that machines exhibit. It can be any kind of understanding and problem solving properties similar to the human brain. Artificial intelligence research includes topics like perception, reasoning, planning, natural language processing (communication) and learning. This book presents the complex subject of artificial intelligence in the most comprehensible and easy to understand language. While understanding the long-term perspectives of the topics, the book makes an effort in highlighting their impact as a modern tool for the growth of the discipline. The topics covered in this extensive text deal with the core subjects of the area. Those in search of information to further their knowledge will be greatly assisted by this textbook.

A detailed account of the significant topics covered in this book is provided below:

Chapter 1- This chapter on artificial intelligence offers an insightful focus, keeping in mind the complex subject matter. Artificial intelligence is the cognitive-like function of machines when they react to a situation or an environment with the goal to maximize their effort at executing a command. There are two approaches taken for defining the scope and view of artificial intelligence. The first way tries to understand human thinking process and builds machines accordingly. The second approach is based on the Turing test.

Chapter 2- Knowledge representation and logic commutes information about the world to a machine in a format that can easily be understood by them. Examples of knowledge representation and logic include systems architecture, ontologies, frames and semantic nets. This section discusses the methods of information representation and logic in a critical manner providing key analysis to the subject matter.

Chapter 3- Knowledge can be represented using logic in the form of facts and rules. However, in complex scenarios, it becomes difficult to do the same. In order to represent complexity, certainty factors are used. Uncertainty comes into play when real world situations are to be evaluated. This chapter elucidates the role and representation of uncertainty.

Chapter 4- Machine learning is an interdisciplinary branch that deals with creating a machine that can learn with experience and not just by programming. Unlike classical artificial intelligence that uses deductive reasoning, this method follows inductive reasoning. While deductive reasoning solves a problem on the basis of general axioms, inductive reasoning takes examples as the base and generates general axioms. The major concepts of machine learning are discussed in this section.

Chapter 5- Natural language processing is a sub-field of computer science that is related with the interaction between computer and human language. Its primary use is to make computers perform tasks when inputs are provided using human language. Linguistic concepts such as syntax, discourse, semantics, etc. are used to evaluate the degree of a task that is to be performed. The topics discussed in the chapter are of great importance to broaden the existing knowledge on natural language processing.

I would like to make a special mention of my publisher who considered me worthy of this opportunity and also supported me throughout the process. I would also like to thank the editing team at the back-end who extended their help whenever required.

Editor

A Comprehensive Overview of Artificial Intelligence

This chapter on artificial intelligence offers an insightful focus, keeping in mind the complex subject matter. Artificial intelligence is the cognitive-like function of machines when they react to a situation or an environment with the goal to maximize their effort at executing a command. There are two approaches taken for defining the scope and view of artificial intelligence. The first way tries to understand human thinking process and builds machines accordingly. The second approach is based on the Turing test.

Artificial Intelligence

Artificial Intelligence is concerned with the design of intelligence in an artificial device.

The term was coined by McCarthy in 1956.

There are two ideas in the definition.

> 1. Intelligence

> 2. Artificial device

What is intelligence?

– Is it that which characterize humans? Or is there an absolute standard of judgement?

– Accordingly there are two possibilities:

> – A system with intelligence is expected to behave as intelligently as a human

> – A system with intelligence is expected to behave in the best possible manner

– Secondly what type of behavior are we talking about?

> – Are we looking at the thought process or reasoning ability of the system?

> – Or are we only interested in the final manifestations of the system in terms of its actions?

Given this scenario different interpretations have been used by different researchers as defining the scope and view of Artificial Intelligence.

> 1. One view is that artificial intelligence is about designing systems that are as intelligent as humans.

This view involves trying to understand human thought and an effort to build machines that emulate the human thought process. This view is the cognitive science approach to AI.

2. The second approach is best embodied by the concept of the Turing Test. Turing held that in future computers can be programmed to acquire abilities rivaling human intelligence. As part of his argument Turing put forward the idea of an 'imitation game', in which a human being and a computer would be interrogated under conditions where the interrogator would not know which was which, the communication being entirely by textual messages. Turing argued that if the interrogator could not distinguish them by questioning, then it would be unreasonable not to call the computer intelligent. Turing's 'imitation game' is now usually called 'the Turing test' for intelligence.

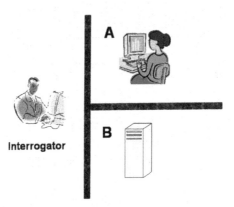

Interrogator

Turing Test

Consider the following setting. There are two rooms, A and B. One of the rooms contains a computer. The other contains a human. The interrogator is outside and does not know which one is a computer. He can ask questions through a teletype and receives answers from both A and B. The interrogator needs to identify whether A or B are humans. To pass the Turing test, the machine has to fool the interrogator into believing that it is human.

3. Logic and laws of thought deals with studies of ideal or rational thought process and inference. The emphasis in this case is on the inferencing mechanism, and its properties. That is how the system arrives at a conclusion, or the reasoning behind its selection of actions is very important in this point of view. The soundness and completeness of the inference mechanisms are important here.

4. The fourth view of AI is that it is the study of rational agents. This view deals with building machines that act rationally. The focus is on how the system acts and performs, and not so much on the reasoning process. A rational agent is one that acts rationally, that is, is in the best possible manner.

Artificial intelligence (AI) is intelligence exhibited by machines. In computer science, the field of AI research defines itself as the study of "intelligent agents": any device that perceives its environment and takes actions that maximize its chance of success at some goal. Colloquially, the term "artificial intelligence" is applied when a machine mimics "cognitive" functions that humans associate with other human minds, such as "learning" and "problem solving" (known as machine learning). As machines become increasingly capable, mental facilities once thought to require intelligence are removed from the definition. For instance, optical character recognition is no longer perceived as an example of "artificial intelligence", having become a routine technology. Capabilities currently classified as AI include successfully understanding human speech, competing at a high level in strategic game systems (such as chess and Go), self-driving cars, intelligent routing in content delivery networks, and interpreting complex data.

AI research is divided into subfields that focus on specific problems or on specific approaches or on the use of a particular tool or towards satisfying particular applications.

The central problems (or goals) of AI research include reasoning, knowledge, planning, learning, natural language processing (communication), perception and the ability to move and manipulate objects. General intelligence is among the field's long-term goals. Approaches include statistical methods, computational intelligence, and traditional symbolic AI. Many tools are used in AI, including versions of search and mathematical optimization, logic, methods based on probability and economics. The AI field draws upon computer science, mathematics, psychology, linguistics, philosophy, neuroscience and artificial psychology.

The field was founded on the claim that human intelligence "can be so precisely described that a machine can be made to simulate it". This raises philosophical arguments about the nature of the mind and the ethics of creating artificial beings endowed with human-like intelligence, issues which have been explored by myth, fiction and philosophy since antiquity. Some people also consider AI a danger to humanity if it progresses unabatedly. Attempts to create artificial intelligence have experienced many setbacks, including the ALPAC report of 1966, the abandonment of perceptrons in 1970, the Lighthill Report of 1973, the second AI winter 1987–1993 and the collapse of the Lisp machine market in 1987.

In the twenty-first century, AI techniques, both "hard" and "soft", have experienced a resurgence following concurrent advances in computer power, sizes of training sets, and theoretical understanding, and AI techniques have become an essential part of the technology industry, helping to solve many challenging problems in computer science.. Recent advancements in AI, and specifically in machine learning, have contributed to the growth of Autonomous Things such as drones and self-driving cars, becoming the main driver of innovation in the automotive industry.

History

While thought-capable artificial beings appeared as storytelling devices in antiquity, the idea of actually trying to build a machine to perform useful reasoning may have begun with Ramon Llull (c. 1300 CE). With his Calculus ratiocinator, Gottfried Leibniz extended the concept of the calculating machine (Wilhelm Schickard engineered the first one around 1623), intending to perform operations on concepts rather than numbers. Since the 19th century, artificial beings are common in fiction, as in Mary Shelley's *Frankenstein* or Karel Čapek's *R.U.R. (Rossum's Universal Robots)*.

The study of mechanical or "formal" reasoning began with philosophers and mathematicians in antiquity. In the 19th century, George Boole refined those ideas into propositional logic and Gottlob Frege developed a notational system for mechanical reasoning (a *"predicate calculus"*).

Around the 1940s, Alan Turing's theory of computation suggested that a machine, by shuffling symbols as simple as "0" and "1", could simulate any conceivable act of mathematical deduction. This insight, that digital computers can simulate any process of formal reasoning, is known as the Church–Turing thesis. Along with concurrent discoveries in neurology, information theory and cybernetics, this led researchers to consider the possibility of building an electronic brain. The first work that is now generally recognized as AI was McCullouch and Pitts' 1943 formal design for Turing-complete "artificial neurons".

The field of AI research was "born" at a conference at Dartmouth College in 1956. Attendees Allen Newell (CMU), Herbert Simon (CMU), John McCarthy (MIT), Marvin Minsky (MIT) and Arthur Samuel (IBM) became the founders and leaders of AI research. At the conference, Newell and Simon, together with programmer J. C. Shaw (RAND), presented the first true artificial intelligence program, the Logic Theorist. This spurred tremendous research in the domain: computers were winning at checkers, solving word problems in algebra, proving logical theorems and speaking English. By the middle of the 1960s, research in the U.S. was heavily funded by the Department of Defense and laboratories had been established around the world. AI's founders were optimistic about the future: Herbert Simon predicted, "machines will be capable, within twenty years, of doing any work a man can do." Marvin Minsky agreed, writing, "within a generation ... the problem of creating 'artificial intelligence' will substantially be solved."

They failed to recognize the difficulty of some of the remaining tasks. Progress slowed and in 1974, in response to the criticism of Sir James Lighthill and ongoing pressure from the US Congress to fund more productive projects, both the U.S. and British governments cut off exploratory research in AI. The next few years would later be called an "AI winter", a period when funding for AI projects was hard to find.

In the early 1980s, AI research was revived by the commercial success of expert systems, a form of AI program that simulated the knowledge and analytical skills of human experts. By 1985 the market for AI had reached over a billion dollars. At the same time, Japan's fifth generation computer project inspired the U.S and British governments to restore funding for academic research. However, beginning with the collapse of the Lisp Machine market in 1987, AI once again fell into disrepute, and a second, longer-lasting hiatus began.

In the late 1990s and early 21st century, AI began to be used for logistics, data mining, medical diagnosis and other areas. The success was due to increasing computational power, greater emphasis on solving specific problems, new ties between AI and other fields and a commitment by researchers to mathematical methods and scientific standards. Deep Blue became the first computer chess-playing system to beat a reigning world chess champion, Garry Kasparov on 11 May 1997.

Advanced statistical techniques (loosely known as deep learning), access to large amounts of data and faster computers enabled advances in machine learning and perception. By the mid 2010s, machine learning applications were used throughout the world. In a *Jeopardy* quiz show exhibition match, IBM's question answering system, Watson, defeated the two greatest Jeopardy champions, Brad Rutter and Ken Jennings, by a significant margin. The Kinect, which provides a 3D body–motion interface for the Xbox 360 and the Xbox One use algorithms that emerged from lengthy AI research as do intelligent personal assistants in smartphones. In March 2016, AlphaGo won 4 out of 5 games of Go in a match with Go champion Lee Sedol, becoming the first computer Go-playing system to beat a professional Go player without handicaps.

According to Bloomberg's Jack Clark, 2015 was a landmark year for artificial intelligence, with the number of software projects that use AI within Google increasing from a "sporadic usage" in 2012 to more than 2,700 projects. Clark also presents factual data indicating that error rates in image processing tasks have fallen significantly since 2011. He attributes this to an increase in affordable neural networks, due to a rise in cloud computing infrastructure and to an increase in research tools and datasets. Other cited examples include Microsoft's development of a Skype system that can automatically translate from one language to another and Facebook's system that can describe images to blind people.

Goals

The overall research goal of artificial intelligence is to create technology that allows computers and machines to function in an intelligent manner. The general problem of simulating (or creating) intelligence has been broken down into sub-problems. These consist of particular traits or capabilities that researchers expect an intelligent system to display. The traits described below have received the most attention.

Erik Sandwell emphasizes planning and learning that is relevant and applicable to the given situation.

Reasoning, Problem Solving

Early researchers developed algorithms that imitated step-by-step reasoning that humans use when they solve puzzles or make logical deductions (reason). By the late 1980s and 1990s, AI research had developed methods for dealing with uncertain or incomplete information, employing concepts from probability and economics.

For difficult problems, algorithms can require enormous computational resources—most experience a "combinatorial explosion": the amount of memory or computer time required becomes astronomical for problems of a certain size. The search for more efficient problem-solving algorithms is a high priority.

Human beings ordinarily use fast, intuitive judgments rather than step-by-step deduction that early AI research was able to model. AI has progressed using "sub-symbolic" problem solving: embodied agent approaches emphasize the importance of sensorimotor skills to higher reasoning; neural net research attempts to simulate the structures inside the brain that give rise to this skill; statistical approaches to AI mimic the human ability.

Knowledge Representation

Knowledge representation and knowledge engineering are central to AI research. Many of the problems machines are expected to solve will require extensive knowledge about the world. Among the things that AI needs to represent are: objects, properties, categories and relations between objects; situations, events, states and time; causes and effects; knowledge about knowledge (what we know about what other people know); and many other, less well researched domains. A representation of "what exists" is an ontology: the set of objects, relations, concepts and so on that the machine knows about. The most general are called upper ontologies, which attempt to provide a foundation for all other knowledge.

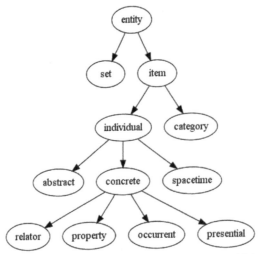

An ontology represents knowledge as a set of concepts within a domain
and the relationships between those concepts.

Among the most difficult problems in knowledge representation are:

Default Reasoning and the Qualification Problem

Many of the things people know take the form of "working assumptions". For example, if a bird comes up in conversation, people typically picture an animal that is fist sized, sings, and flies. None of these things are true about all birds. John McCarthy identified this problem in 1969 as the qualification problem: for any commonsense rule that AI researchers care to represent, there tend to be a huge number of exceptions. Almost nothing is simply true or false in the way that abstract logic requires. AI research has explored a number of solutions to this problem.

The Breadth of Commonsense Knowledge

The number of atomic facts that the average person knows is very large. Research projects that attempt to build a complete knowledge base of commonsense knowledge (e.g., Cyc) require enormous amounts of laborious ontological engineering—they must be built, by hand, one complicated concept at a time. A major goal is to have the computer understand enough concepts to be able to learn by reading from sources like the Internet, and thus be able to add to its own ontology.

The Subsymbolic form of Some Commonsense Knowledge

Much of what people know is not represented as "facts" or "statements" that they could express verbally. For example, a chess master will avoid a particular chess position because it "feels too exposed" or an art critic can take one look at a statue and realize that it is a fake. These are intuitions or tendencies that are represented in the brain non-consciously and sub-symbolically. Knowledge like this informs, supports and provides a context for symbolic, conscious knowledge. As with the related problem of sub-symbolic reasoning, it is hoped that situated AI, computational intelligence, or statistical AI will provide ways to represent this kind of knowledge.

Planning

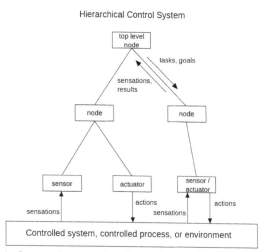

A hierarchical control system is a form of control system in which a set of devices and governing software is arranged in a hierarchy.

Intelligent agents must be able to set goals and achieve them. They need a way to visualize the future (they must have a representation of the state of the world and be able to make predictions about how their actions will change it) and be able to make choices that maximize the utility (or "value") of the available choices.

In classical planning problems, the agent can assume that it is the only thing acting on the world and it can be certain what the consequences of its actions may be. However, if the agent is not the only actor, it must periodically ascertain whether the world matches its predictions and it must change its plan as this becomes necessary, requiring the agent to reason under uncertainty.

Multi-agent planning uses the cooperation and competition of many agents to achieve a given goal. Emergent behavior such as this is used by evolutionary algorithms and swarm intelligence.

Learning

Machine learning is the study of computer algorithms that improve automatically through experience and has been central to AI research since the field's inception.

Unsupervised learning is the ability to find patterns in a stream of input. Supervised learning includes both classification and numerical regression. Classification is used to determine what category something belongs in, after seeing a number of examples of things from several categories. Regression is the attempt to produce a function that describes the relationship between inputs and outputs and predicts how the outputs should change as the inputs change. In reinforcement learning the agent is rewarded for good responses and punished for bad ones. The agent uses this sequence of rewards and punishments to form a strategy for operating in its problem space. These three types of learning can be analyzed in terms of decision theory, using concepts like utility. The mathematical analysis of machine learning algorithms and their performance is a branch of theoretical computer science known as computational learning theory.

Within developmental robotics, developmental learning approaches were elaborated for lifelong cumulative acquisition of repertoires of novel skills by a robot, through autonomous self-exploration and social interaction with human teachers, and using guidance mechanisms such as active learning, maturation, motor synergies, and imitation.

Natural Language Processing

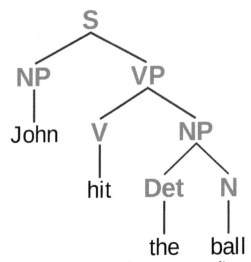

A parse tree represents the syntactic structure of a sentence according to some formal grammar.

Natural language processing gives machines the ability to read and understand the languages that humans speak. A sufficiently powerful natural language processing system would enable natural language user interfaces and the acquisition of knowledge directly from human-written sources, such as newswire texts. Some straightforward applications of natural language processing include information retrieval, text mining, question answering and machine translation.

A common method of processing and extracting meaning from natural language is through semantic indexing. Increases in processing speeds and the drop in the cost of data storage makes indexing large volumes of abstractions of the user's input much more efficient.

Perception

Machine perception is the ability to use input from sensors (such as cameras, microphones, tactile sensors, sonar and others more exotic) to deduce aspects of the world. Computer vision is the ability to analyze visual input. A few selected subproblems are speech recognition, facial recognition and object recognition.

Motion and Manipulation

The field of robotics is closely related to AI. Intelligence is required for robots to be able to handle such tasks as object manipulation and navigation, with sub-problems of localization (knowing where you are, or finding out where other things are), mapping (learning what is around you, building a map of the environment), and motion planning (figuring out how to get there) or path planning (going from one point in space to another point, which may involve compliant motion – where the robot moves while maintaining physical contact with an object).

Social Intelligence

Kismet, a robot with rudimentary social skills

Affective computing is the study and development of systems and devices that can recognize, interpret, process, and simulate human affects. It is an interdisciplinary field spanning computer sciences, psychology, and cognitive science. While the origins of the field may be traced as far back as to early philosophical inquiries into emotion, the more modern branch of computer science originated with Rosalind Picard's 1995 paper on affective computing. A motivation for the research is the ability to simulate empathy. The machine should interpret the emotional state of humans and adapt its behaviour to them, giving an appropriate response for those emotions.

Emotion and social skills play two roles for an intelligent agent. First, it must be able to predict the actions of others, by understanding their motives and emotional states. (This involves elements of game theory, decision theory, as well as the ability to model human emotions and the perceptual skills to detect emotions.) Also, in an effort to facilitate human-computer interaction, an intelligent machine might want to be able to *display* emotions—even if it does not actually experience them itself—in order to appear sensitive to the emotional dynamics of human interaction.

Creativity

A sub-field of AI addresses creativity both theoretically (from a philosophical and psychological perspective) and practically (via specific implementations of systems that generate outputs that can be considered creative, or systems that identify and assess creativity). Related areas of computational research are Artificial intuition and Artificial thinking.

General Intelligence

Many researchers think that their work will eventually be incorporated into a machine with artificial general intelligence, combining all the skills above and exceeding human abilities at most or all of them. A few believe that anthropomorphic features like artificial consciousness or an artificial brain may be required for such a project.

Many of the problems above may require general intelligence to be considered solved. For example, even a straightforward, specific task like machine translation requires that the machine read and write in both languages (NLP), follow the author's argument (reason), know what is being talked about (knowledge), and faithfully reproduce the author's intention (social intelligence). A problem like machine translation is considered "AI-complete". In order to reach human-level performance for machines, one must solve all the problems.

Approaches

There is no established unifying theory or paradigm that guides AI research. Researchers disagree about many issues. A few of the most long standing questions that have remained unanswered are these: should artificial intelligence simulate natural intelligence by studying psychology or neurology? Or is human biology as irrelevant to AI research as bird biology is to aeronautical engineering? Can intelligent behavior be described using simple, elegant principles (such as logic or optimization)? Or does it necessarily require solving a large number of completely unrelated problems? Can intelligence be reproduced using high-level symbols, similar to words and ideas? Or does it require "sub-symbolic" processing? John Haugeland, who coined the term GOFAI (Good Old-Fashioned Artificial Intelligence), also proposed that AI should more properly be referred to as synthetic intelligence, a term which has since been adopted by some non-GOFAI researchers.

Stuart Shapiro divides AI research into three approaches, which he calls computational psychology, computational philosophy, and computer science. Computational psychology is used to make computer programs that mimic human behavior. Computational philosophy, is used to develop an adaptive, free-flowing computer mind. Implementing computer science serves the goal of creating computers that can perform tasks that only people could previously accomplish. Together, the humanesque behavior, mind, and actions make up artificial intelligence.

Cybernetics and Brain Simulation

In the 1940s and 1950s, a number of researchers explored the connection between neurology, information theory, and cybernetics. Some of them built machines that used electronic networks to exhibit rudimentary intelligence, such as W. Grey Walter's turtles and the Johns Hopkins Beast. Many of these researchers gathered for meetings of the Teleological Society at Princeton University and the Ratio Club in England. By 1960, this approach was largely abandoned, although elements of it would be revived in the 1980s.

Symbolic

When access to digital computers became possible in the middle 1950s, AI research began to explore the possibility that human intelligence could be reduced to symbol manipulation. The research was centered in three institutions: Carnegie Mellon University, Stanford and MIT, and each one developed its own style of research. John Haugeland named these approaches to AI "good old fashioned AI" or "GOFAI". During the 1960s, symbolic approaches had achieved great success at simulating high-level thinking in small demonstration programs. Approaches based on cybernetics or neural networks were abandoned or pushed into the background. Researchers in the 1960s and the 1970s were convinced that symbolic approaches would eventually succeed in creating a machine with artificial general intelligence and considered this the goal of their field.

Cognitive Simulation

Economist Herbert Simon and Allen Newell studied human problem-solving skills and attempted to formalize them, and their work laid the foundations of the field of artificial intelligence, as well as cognitive science, operations research and management science. Their research team used the results of psychological experiments to develop programs that simulated the techniques that people used to solve problems. This tradition, centered at Carnegie Mellon University would eventually culminate in the development of the Soar architecture in the middle 1980s.

Logic-based

Unlike Newell and Simon, John McCarthy felt that machines did not need to simulate human thought, but should instead try to find the essence of abstract reasoning and problem solving, regardless of whether people used the same algorithms. His laboratory at Stanford (SAIL) focused on using formal logic to solve a wide variety of problems, including knowledge representation, planning and learning. Logic was also the focus of the work at the University of Edinburgh and elsewhere in Europe which led to the development of the programming language Prolog and the science of logic programming.

Anti-logic or Scruffy

Researchers at MIT (such as Marvin Minsky and Seymour Papert) found that solving difficult problems in vision and natural language processing required ad-hoc solutions – they argued that there was no simple and general principle (like logic) that would capture all the aspects of intelligent behavior. Roger Schank described their "anti-logic" approaches as "scruffy" (as opposed to the "neat" paradigms at CMU and Stanford). Commonsense knowledge bases (such as Doug Lenat's Cyc) are an example of "scruffy" AI, since they must be built by hand, one complicated concept at a time.

Knowledge-based

When computers with large memories became available around 1970, researchers from all three traditions began to build knowledge into AI applications. This "knowledge revolution" led to the development and deployment of expert systems (introduced by Edward Feigenbaum), the first truly successful form of AI software. The knowledge revolution was also driven by the realization that enormous amounts of knowledge would be required by many simple AI applications.

Sub-symbolic

By the 1980s progress in symbolic AI seemed to stall and many believed that symbolic systems would never be able to imitate all the processes of human cognition, especially perception, robotics, learning and pattern recognition. A number of researchers began to look into "sub-symbolic" approaches to specific AI problems. Sub-symbolic methods manage to approach intelligence without specific representations of knowledge.

Bottom-up, Embodied, Situated, Behavior-based or Nouvelle AI

Researchers from the related field of robotics, such as Rodney Brooks, rejected symbolic AI and focused on the basic engineering problems that would allow robots to move and survive. Their work revived the non-symbolic viewpoint of the early cybernetics researchers of the 1950s and reintroduced the use of control theory in AI. This coincided with the development of the embodied mind thesis in the related field of cognitive science: the idea that aspects of the body (such as movement, perception and visualization) are required for higher intelligence.

Computational Intelligence and Soft Computing

Interest in neural networks and "connectionism" was revived by David Rumelhart and others in the middle of 1980s. Neural networks are an example of soft computing --- they are solutions to problems which cannot be solved with complete logical certainty, and where an approximate solution is often sufficient. Other soft computing approaches to AI include fuzzy systems, evolutionary computation and many statistical tools. The application of soft computing to AI is studied collectively by the emerging discipline of computational intelligence.

Statistical

In the 1990s, AI researchers developed sophisticated mathematical tools to solve specific sub-problems. These tools are truly scientific, in the sense that their results are both measurable and verifiable, and they have been responsible for many of AI's recent successes. The shared mathematical language has also permitted a high level of collaboration with more established fields (like mathematics, economics or operations research). Stuart Russell and Peter Norvig describe this movement as nothing less than a "revolution" and "the victory of the neats". Critics argue that these techniques (with few exceptions) are too focused on particular problems and have failed to address the long-term goal of general intelligence. There is an ongoing debate about the relevance and validity of statistical approaches in AI, exemplified in part by exchanges between Peter Norvig and Noam Chomsky.

Integrating the approaches

Intelligent Agent Paradigm

An intelligent agent is a system that perceives its environment and takes actions which maximize its chances of success. The simplest intelligent agents are programs that solve specific problems. More complicated agents include human beings and organizations of human beings (such as firms). The paradigm gives researchers license to study isolated problems and find solutions that are both verifiable and useful, without agreeing on one single approach. An agent that solves a specific problem can use any approach that works – some agents are symbolic and logical, some are sub-symbolic neural networks and others may use new approaches. The paradigm also gives researchers a common language to communicate with other fields—such as decision theory and economics—that also use concepts of abstract agents. The intelligent agent paradigm became widely accepted during the 1990s.

Agent Architectures and Cognitive Architectures

Researchers have designed systems to build intelligent systems out of interacting intelligent agents in a multi-agent system. A system with both symbolic and sub-symbolic components is a hybrid intelligent system, and the study of such systems is artificial intelligence systems integration. A hierarchical control system provides a bridge between sub-symbolic AI at its lowest, reactive levels and traditional symbolic AI at its highest levels, where relaxed time constraints permit planning and world modelling. Rodney Brooks' subsumption architecture was an early proposal for such a hierarchical system.

Tools

In the course of 50 years of research, AI has developed a large number of tools to solve the most difficult problems in computer science. A few of the most general of these methods are discussed below.

Search and Optimization

Many problems in AI can be solved in theory by intelligently searching through many possible solutions: Reasoning can be reduced to performing a search. For example, logical proof can be viewed as searching for a path that leads from premises to conclusions, where each step is the application of an inference rule. Planning algorithms search through trees of goals and subgoals, attempting to find a path to a target goal, a process called means-ends analysis. Robotics algorithms for moving limbs and grasping objects use local searches in configuration space. Many learning algorithms use search algorithms based on optimization.

Simple exhaustive searches are rarely sufficient for most real world problems: the search space (the number of places to search) quickly grows to astronomical numbers. The result is a search that is too slow or never completes. The solution, for many problems, is to use "heuristics" or "rules of thumb" that eliminate choices that are unlikely to lead to the goal (called "pruning the search tree"). Heuristics supply the program with a "best guess" for the path on which the solution lies. Heuristics limit the search for solutions into a smaller sample size.

A very different kind of search came to prominence in the 1990s, based on the mathematical theory of optimization. For many problems, it is possible to begin the search with some form of a guess and then refine the guess incrementally until no more refinements can be made. These algorithms can be visualized as blind hill climbing: we begin the search at a random point on the landscape, and then, by jumps or steps, we keep moving our guess uphill, until we reach the top. Other optimization algorithms are simulated annealing, beam search and random optimization.

Evolutionary computation uses a form of optimization search. For example, they may begin with a population of organisms (the guesses) and then allow them to mutate and recombine, selecting only the fittest to survive each generation (refining the guesses). Forms of evolutionary computation include swarm intelligence algorithms (such as ant colony or particle swarm optimization) and evolutionary algorithms (such as genetic algorithms, gene expression programming, and genetic programming).

Logic

Logic is used for knowledge representation and problem solving, but it can be applied to other problems as well. For example, the satplan algorithm uses logic for planning and inductive logic programming is a method for learning.

Several different forms of logic are used in AI research. Propositional or sentential logic is the logic of statements which can be true or false. First-order logic also allows the use of quantifiers and predicates, and can express facts about objects, their properties, and their relations with each other. Fuzzy logic, is a version of first-order logic which allows the truth of a statement to be represented as a value between 0 and 1, rather than simply True (1) or False (0). Fuzzy systems can be used for uncertain reasoning and have been widely used in modern industrial and consumer product control systems. Subjective logic models uncertainty in a different and more explicit manner than fuzzy-logic: a given binomial opinion satisfies belief + disbelief + uncertainty = 1 within a Beta distribution. By this method, ignorance can be distinguished from probabilistic statements that an agent makes with high confidence.

Default logics, non-monotonic logics and circumscription are forms of logic designed to help with default reasoning and the qualification problem. Several extensions of logic have been designed to handle specific domains of knowledge, such as: description logics; situation calculus, event calculus and fluent calculus (for representing events and time); causal calculus; belief calculus; and modal logics.

Probabilistic Methods for Uncertain Reasoning

Many problems in AI (in reasoning, planning, learning, perception and robotics) require the agent to operate with incomplete or uncertain information. AI researchers have devised a number of powerful tools to solve these problems using methods from probability theory and economics.

Bayesian networks are a very general tool that can be used for a large number of problems: reasoning (using the Bayesian inference algorithm), learning (using the expectation-maximization algorithm), planning (using decision networks) and perception (using dynamic Bayesian networks). Probabilistic algorithms can also be used for filtering, prediction, smoothing and finding explanations for streams of data, helping perception systems to analyze processes that occur over time (e.g., hidden Markov models or Kalman filters).

A key concept from the science of economics is "utility": a measure of how valuable something is to an intelligent agent. Precise mathematical tools have been developed that analyze how an agent can make choices and plan, using decision theory, decision analysis, and information value theory. These tools include models such as Markov decision processes, dynamic decision networks, game theory and mechanism design.

Classifiers and Statistical Learning Methods

The simplest AI applications can be divided into two types: classifiers ("if shiny then diamond") and controllers ("if shiny then pick up"). Controllers do, however, also classify conditions before inferring actions, and therefore classification forms a central part of many AI systems. Classifiers are functions that use pattern matching to determine a closest match. They can be tuned according to examples, making them very attractive for use in AI. These examples are known as observations or patterns. In supervised learning, each pattern belongs to a certain predefined class. A class can be seen as a decision that has to be made. All the observations combined with their class labels are known as a data set. When a new observation is received, that observation is classified based on previous experience.

A classifier can be trained in various ways; there are many statistical and machine learning approaches. The most widely used classifiers are the neural network, kernel methods such as the support vector machine, k-nearest neighbor algorithm, Gaussian mixture model, naive Bayes classifier, and decision tree. The performance of these classifiers have been compared over a wide range of tasks. Classifier performance depends greatly on the characteristics of the data to be classified. There is no single classifier that works best on all given problems; this is also referred to as the "no free lunch" theorem. Determining a suitable classifier for a given problem is still more an art than science.

Neural Networks

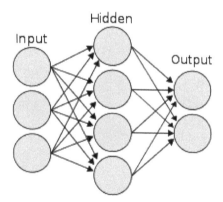

A neural network is an interconnected group of nodes, akin to the vast network of neurons in the human brain.

The study of non-learning artificial neural networks began in the decade before the field of AI research was founded, in the work of Walter Pitts and Warren McCullouch. Frank Rosenblatt invented the perceptron, a learning network with a single layer, similar to the old concept of linear regression. Early pioneers also include Alexey Grigorevich Ivakhnenko, Teuvo Kohonen, Stephen Grossberg, Kunihiko Fukushima, Christoph von der Malsburg, David Willshaw, Shun-Ichi Amari, Bernard Widrow, John Hopfield, Eduardo R. Caianiello, and others.

The main categories of networks are acyclic or feedforward neural networks (where the signal passes in only one direction) and recurrent neural networks (which allow feedback and short-term memories of previous input events). Among the most popular feedforward networks are perceptrons, multi-layer perceptrons and radial basis networks. Neural networks can be applied to the problem of intelligent control (for robotics) or learning, using such techniques as Hebbian learning, GMDH or competitive learning.

Today, neural networks are often trained by the backpropagation algorithm, which had been around since 1970 as the reverse mode of automatic differentiation published by Seppo Linnain-maa, and was introduced to neural networks by Paul Werbos.

Hierarchical temporal memory is an approach that models some of the structural and algorithmic properties of the neocortex.

Deep Feedforward Neural Networks

Deep learning in artificial neural networks with many layers has transformed many important subfields of artificial intelligence, including computer vision, speech recognition, natural language processing and others.

According to a survey, the expression "Deep Learning" was introduced to the Machine Learning community by Rina Dechter in 1986 and gained traction after Igor Aizenberg and colleagues introduced it to Artificial Neural Networks in 2000. The first functional Deep Learning networks were published by Alexey Grigorevich Ivakhnenko and V. G. Lapa in 1965. These networks are trained one layer at a time. Ivakhnenko's 1971 paper describes the learning of a deep feedforward multilayer perceptron with eight layers, already much deeper than many later networks. In 2006, a publication by Geoffrey Hinton and Ruslan Salakhutdinov introduced another way of pre-training many-layered feedforward neural networks (FNNs) one layer at a time, treating each layer in turn as an unsupervised restricted Boltzmann machine, then using supervised backpropagation for fine-tuning. Similar to shallow artificial neural networks, deep neural networks can model complex non-linear relationships. Over the last few years, advances in both machine learning algorithms and computer hardware have led to more efficient methods for training deep neural networks that contain many layers of non-linear hidden units and a very large output layer.

Deep learning often uses convolutional neural networks (CNNs), whose origins can be traced back to the Neocognitron introduced by Kunihiko Fukushima in 1980. In 1989, Yann LeCun and colleagues applied backpropagation to such an architecture. In the early 2000s, in an industrial application CNNs already processed an estimated 10% to 20% of all the checks written in the US. Since 2011, fast implementations of CNNs on GPUs have won many visual pattern recognition competitions.

Deep feedforward neural networks were used in conjunction with reinforcement learning by AlphaGo, Google Deepmind's program that was the first to beat a professional human player.

Deep Recurrent Neural Networks

Early on, deep learning was also applied to sequence learning with recurrent neural networks (RNNs) which are general computers and can run arbitrary programs to process arbitrary sequences of inputs. The depth of an RNN is unlimited and depends on the length of its input sequence. RNNs can be trained by gradient descent but suffer from the vanishing gradient problem. In 1992, it was shown that unsupervised pre-training of a stack of recurrent neural networks can speed up subsequent supervised learning of deep sequential problems.

Numerous researchers now use variants of a deep learning recurrent NN called the long short-term memory (LSTM) network published by Hochreiter & Schmidhuber in 1997. LSTM is often trained by Connectionist Temporal Classification (CTC). At Google, Microsoft and Baidu this approach has revolutionised speech recognition. For example, in 2015, Google's speech recognition experienced a dramatic performance jump of 49% through CTC-trained LSTM, which is now available through Google Voice to billions of smartphone users. Google also used LSTM to improve machine translation, Language Modeling and Multilingual Language Processing. LSTM combined with CNNs also improved automatic image captioning and a plethora of other applications.

Control Theory

Control theory, the grandchild of cybernetics, has many important applications, especially in robotics.

Languages

AI researchers have developed several specialized languages for AI research, including Lisp and Prolog.

Evaluating Progress

In 1950, Alan Turing proposed a general procedure to test the intelligence of an agent now known as the Turing test. This procedure allows almost all the major problems of artificial intelligence to be tested. However, it is a very difficult challenge and at present all agents fail.

Artificial intelligence can also be evaluated on specific problems such as small problems in chemistry, hand-writing recognition and game-playing. Such tests have been termed subject matter expert Turing tests. Smaller problems provide more achievable goals and there are an ever-increasing number of positive results.

For example, performance at draughts (i.e. checkers) is optimal, performance at chess is high-human and nearing super-human and performance at many everyday tasks (such as recognizing a face or crossing a room without bumping into something) is sub-human.

A quite different approach measures machine intelligence through tests which are developed from *mathematical* definitions of intelligence. Examples of these kinds of tests start in the late nineties devising intelligence tests using notions from Kolmogorov complexity and data compression. Two major advantages of mathematical definitions are their applicability to nonhuman intelligences and their absence of a requirement for human testers.

A derivative of the Turing test is the Completely Automated Public Turing test to tell Computers and Humans Apart (CAPTCHA). As the name implies, this helps to determine that a user is an actual person and not a computer posing as a human. In contrast to the standard Turing test, CAPTCHA administered by a machine and targeted to a human as opposed to being administered by a human and targeted to a machine. A computer asks a user to complete a simple test then generates a grade for that test. Computers are unable to solve the problem, so correct solutions are deemed to be the result of a person taking the test. A common type of CAPTCHA is the test that requires the typing of distorted letters, numbers or symbols that appear in an image undecipherable by a computer.

Applications

AI is relevant to any intellectual task. Modern artificial intelligence techniques are pervasive and are too numerous to list here. Frequently, when a technique reaches mainstream use, it is no longer considered artificial intelligence; this phenomenon is described as the AI effect.

An automated online assistant providing customer service
on a web page – one of many very primitive applications of artificial intelligence.

High-profile examples of AI include autonomous vehicles (such as drones and self-driving cars), medical diagnosis, creating art (such as poetry), proving mathematical theorems, playing games (such as Chess or Go), search engines (such as Google search), online assistants (such as Siri), image recognition in photographs, spam filtering, prediction of judicial decisions and targeting online advertisements.

With social media sites overtaking TV as a source for news for young people and news organisations increasingly reliant on social media platforms for generating distribution, major publishers now use artificial intelligence (AI) technology to post stories more effectively and generate higher volumes of traffic.

Competitions and Prizes

There are a number of competitions and prizes to promote research in artificial intelligence. The main areas promoted are: general machine intelligence, conversational behavior, data-mining, robotic cars, robot soccer and games.

Healthcare

Artificial intelligence is breaking into the healthcare industry by assisting doctors. According to Bloomberg Technology, Microsoft has developed AI to help doctors find the right treatments for cancer. There is a great amount of research and drugs developed relating to cancer. In detail, there are more than 800 medicines and vaccines to treat cancer. This negatively affects the doctors, because there are way too many options to choose from, making it more difficult to choose the right drugs for the patients. Microsoft is working on a project to develop a machine called "Hanover". Its goal is to memorize all the papers necessary to cancer and help predict which combinations of drugs will be most effective for each patient. One project that is being worked on at the moment is fighting myeloid leukemia, a fatal cancer where the treatment has not improved in decades. Another study was reported to have found that artificial intelligence was as good as trained doctors in identifying skin cancers. Another study is using artificial intelligence to try and monitor multiple high-risk patients, and this is done by asking each patient numerous questions based on data acquired from live doctor to patient interactions.

According to CNN, there was a recent study by surgeons at the Children's National Medical Center in Washington which successfully demonstrated surgery with an autonomous robot. The team supervised the robot while it performed soft-tissue surgery, stitching together a pig's bowel during open surgery, and doing so better than a human surgeon, the team claimed.

Automotive Industry

Advancements in AI have contributed to the growth of the automotive industry through the creation and evolution of self-driving vehicles. As of 2016, there are over 30 companies utilizing AI into the creation of driverless cars. A few companies involved with AI include Tesla, Google, and Apple.

Many components contribute to the functioning of self-driving cars. These vehicles incorporate systems such as braking, lane changing, collision prevention, navigation and mapping. Together, these systems, as well as high performance computers are integrated into one complex vehicle.

One main factor that influences the ability for a driver-less car to function is mapping. In general, the vehicle would be pre-programmed with a map of the area being driven. This map would include data on the approximations of street light and curb heights in order for the vehicle to be aware of its surroundings. However, Google has been working on an algorithm with the purpose of eliminating the need for pre-programmed maps and instead, creating a device that would be able to adjust to a variety of new surroundings. Some self-driving cars are not equipped with steering wheels or brakes, so there has also been research focused on creating an algorithm that is capable of maintaining a safe environment for the passengers in the vehicle through awareness of speed and driving conditions.

Finance

Financial institutions have long used artificial neural network systems to detect charges or claims outside of the norm, flagging these for human investigation.

Use of AI in banking can be tracked back to 1987 when Security Pacific National Bank in USA setup a Fraud Prevention Task force to counter the unauthorised use of debit cards. Apps like Kasisito and Moneystream are using AI in financial services Banks use artificial intelligence systems to organize operations, maintain book-keeping, invest in stocks, and manage properties. AI can react to changes overnight or when business is not taking place. In August 2001, robots beat humans in a simulated financial trading competition.

AI has also reduced fraud and crime by monitoring behavioral patterns of users for any changes or anomalies.

Platforms

A platform (or "computing platform") is defined as "some sort of hardware architecture or software framework (including application frameworks), that allows software to run". As Rodney Brooks pointed out many years ago, it is not just the artificial intelligence software that defines the AI features of the platform, but rather the actual platform itself that affects the AI that results, i.e., there needs to be work in AI problems on real-world platforms rather than in isolation.

A wide variety of platforms has allowed different aspects of AI to develop, ranging from expert systems such as Cyc to deep-learning frameworks to robot platforms such as the Roomba with open interface. Recent advances in deep artificial neural networks and distributed computing have led to a proliferation of software libraries, including Deeplearning4j, TensorFlow, Theano and Torch.

Partnership on AI

Amazon, Google, Facebook, IBM, and Microsoft have established a non-profit partnership to formulate best practices on artificial intelligence technologies, advance the public's understanding, and to serve as a platform about artificial intelligence. They stated: "This partnership on AI will conduct research, organize discussions, provide thought leadership, consult with relevant third parties, respond to questions from the public and media, and create educational material that advance the understanding of AI technologies including machine perception, learning, and automated reasoning." Apple joined other tech companies as a founding member of the Partnership on AI in January 2017. The corporate members will make financial and research contributions to the group, while engaging with the scientific community to bring academics onto the board.

Philosophy and Ethics

There are three philosophical questions related to AI:

1. Is artificial general intelligence possible? Can a machine solve any problem that a human being can solve using intelligence? Or are there hard limits to what a machine can accomplish?

2. Are intelligent machines dangerous? How can we ensure that machines behave ethically and that they are used ethically?

3. Can a machine have a mind, consciousness and mental states in exactly the same sense that human beings do? Can a machine be sentient, and thus deserve certain rights? Can a machine intentionally cause harm?

The limits of artificial general intelligence

Can a machine be intelligent? Can it "think"?

Turing's "Polite Convention"

We need not decide if a machine can "think"; we need only decide if a machine can act as intelligently as a human being. This approach to the philosophical problems associated with artificial intelligence forms the basis of the Turing test.

The Dartmouth Proposal

"Every aspect of learning or any other feature of intelligence can be so precisely described that a machine can be made to simulate it." This conjecture was printed in the proposal for the Dartmouth Conference of 1956, and represents the position of most working AI researchers.

Newell and Simon's Physical Symbol System Hypothesis

"A physical symbol system has the necessary and sufficient means of general intelligent action." Newell and Simon argue that intelligence consists of formal operations on symbols. Hubert Dreyfus argued that, on the contrary, human expertise depends on unconscious instinct rather than conscious symbol manipulation and on having a "feel" for the situation rather than explicit symbolic knowledge.

Gödelian Arguments

Gödel himself, John Lucas (in 1961) and Roger Penrose (in a more detailed argument from 1989 onwards) made highly technical arguments that human mathematicians can consistently see the truth of their own "Gödel statements" and therefore have computational abilities beyond that of mechanical Turing machines. However, the modern consensus in the scientific and mathematical community is that these "Gödelian arguments" fail.

The Artificial Brain Argument

The brain can be simulated by machines and because brains are intelligent, simulated brains must also be intelligent; thus machines can be intelligent. Hans Moravec, Ray Kurzweil and others have argued that it is technologically feasible to copy the brain directly into hardware and software, and that such a simulation will be essentially identical to the original.

The AI effect

Machines are *already* intelligent, but observers have failed to recognize it. When Deep Blue beat Garry Kasparov in chess, the machine was acting intelligently. However, onlookers commonly discount the behavior of an artificial intelligence program by arguing that it is not "real" intelligence after all; thus "real" intelligence is whatever intelligent behavior people can do that machines still can not. This is known as the AI Effect: "AI is whatever hasn't been done yet."

Potential Risks and Moral Reasoning

Widespread use of artificial intelligence could have unintended consequences that are dangerous or undesirable. Scientists from the Future of Life Institute, among others, described some short-term research goals to be how AI influences the economy, the laws and ethics that are involved with AI and how to minimize AI security risks. In the long-term, the scientists have proposed to continue optimizing function while minimizing possible security risks that come along with new technologies.

Machines with intelligence have the potential to use their intelligence to make ethical decisions. Research in this area includes "machine ethics", "artificial moral agents", and the study of "malevolent vs. friendly AI".

Existential Risk

The development of full artificial intelligence could spell the end of the human race. Once humans develop artificial intelligence, it will take off on its own and redesign itself at an ever-increasing rate. Humans, who are limited by slow biological evolution, couldn't compete and would be superseded.

—Stephen Hawking

A common concern about the development of artificial intelligence is the potential threat it could pose to mankind. This concern has recently gained attention after mentions by celebrities including Stephen Hawking, Bill Gates, and Elon Musk. A group of prominent tech titans including Peter Thiel, Amazon Web Services and Musk have committed $1billion to OpenAI a nonprofit company aimed at championing responsible AI development. The opinion of experts within the field of artificial intelligence is mixed, with sizable fractions both concerned and unconcerned by risk from eventual superhumanly-capable AI.

In his book *Superintelligence*, Nick Bostrom provides an argument that artificial intelligence will pose a threat to mankind. He argues that sufficiently intelligent AI, if it chooses actions based on achieving some goal, will exhibit convergent behavior such as acquiring resources or protecting itself from being shut down. If this AI's goals do not reflect humanity's - one example is an AI told to compute as many digits of pi as possible - it might harm humanity in order to acquire more resources or prevent itself from being shut down, ultimately to better achieve its goal.

For this danger to be realized, the hypothetical AI would have to overpower or out-think all of humanity, which a minority of experts argue is a possibility far enough in the future to not be worth researching. Other counterarguments revolve around humans being either intrinsically or convergently valuable from the perspective of an artificial intelligence.

Concern over risk from artificial intelligence has led to some high-profile donations and investments. In January 2015, Elon Musk donated ten million dollars to the Future of Life Institute to fund research on understanding AI decision making. The goal of the institute is to "grow wisdom with which we manage" the growing power of technology. Musk also funds companies developing artificial intelligence such as Google DeepMind and Vicarious to "just keep an eye on what's going on with artificial intelligence. I think there is potentially a dangerous outcome there."

Development of militarized artificial intelligence is a related concern. Currently, 50+ countries are researching battlefield robots, including the United States, China, Russia, and the United Kingdom. Many people concerned about risk from superintelligent AI also want to limit the use of artificial soldiers.

Devaluation of Humanity

Joseph Weizenbaum wrote that AI applications can not, by definition, successfully simulate genuine human empathy and that the use of AI technology in fields such as customer service or psychotherapy was deeply misguided. Weizenbaum was also bothered that AI researchers (and some philosophers) were willing to view the human mind as nothing more than a computer program (a position now known as computationalism). To Weizenbaum these points suggest that AI research devalues human life.

Decrease in Demand for Human Labor

Martin Ford, author of *The Lights in the Tunnel: Automation, Accelerating Technology and the Economy of the Future*, and others argue that specialized artificial intelligence applications, robotics and other forms of automation will ultimately result in significant unemployment as machines begin to match and exceed the capability of workers to perform most routine and repetitive jobs. Ford predicts that many knowledge-based occupations—and in particular entry level jobs—will be increasingly susceptible to automation via expert systems, machine learning and other AI-enhanced applications. AI-based applications may also be used to amplify the capabilities of low-wage offshore workers, making it more feasible to outsource knowledge work.

Artificial Moral Agents

This raises the issue of how ethically the machine should behave towards both humans and other AI agents. This issue was addressed by Wendell Wallach in his book titled *Moral Machines* in which he introduced the concept of artificial moral agents (AMA). For Wallach, AMAs have become a part of the research landscape of artificial intelligence as guided by its two central questions which he identifies as "Does Humanity Want Computers Making Moral Decisions" and "Can (Ro) bots Really Be Moral". For Wallach the question is not centered on the issue of *whether* machines can demonstrate the equivalent of moral behavior in contrast to the *constraints* which society may place on the development of AMAs.

Machine Ethics

The field of machine ethics is concerned with giving machines ethical principles, or a procedure for discovering a way to resolve the ethical dilemmas they might encounter, enabling them to function in an ethically responsible manner through their own ethical decision making. The field was delineated in the AAAI Fall 2005 Symposium on Machine Ethics: "Past research concerning the relationship between technology and ethics has largely focused on responsible and irresponsible use of technology by human beings, with a few people being interested in how human beings ought to treat machines. In all cases, only human beings have engaged in ethical reasoning. The time has come for adding an ethical dimension to at least some machines. Recognition of the ethical ramifications of behavior involving machines, as well as recent and potential developments in machine autonomy, necessitate this. In contrast to computer hacking, software property issues, privacy issues and other topics normally ascribed to computer ethics, machine ethics is concerned with the behavior of machines towards human users and other machines. Research in machine ethics is key to alleviating concerns with autonomous systems—it could be argued that the notion of autonomous machines without such a dimension is at the root of all fear concerning machine intelligence. Further, investigation of machine ethics could enable the discovery of problems with current ethical theories, advancing our thinking about Ethics." Machine ethics is sometimes referred to as machine morality, computational ethics or computational morality. A variety of perspectives of this nascent field can be found in the collected edition "Machine Ethics" that stems from the AAAI Fall 2005 Symposium on Machine Ethics.

Malevolent and Friendly AI

Political scientist Charles T. Rubin believes that AI can be neither designed nor guaranteed to be benevolent. He argues that "any sufficiently advanced benevolence may be indistinguishable from malevolence." Humans should not assume machines or robots would treat us favorably, because there is no *a priori* reason to believe that they would be sympathetic to our system of morality, which has evolved along with our particular biology (which AIs would not share). Hyper-intelligent software may not necessarily decide to support the continued existence of mankind, and would be extremely difficult to stop. This topic has also recently begun to be discussed in academic publications as a real source of risks to civilization, humans, and planet Earth.

Physicist Stephen Hawking, Microsoft founder Bill Gates and SpaceX founder Elon Musk have expressed concerns about the possibility that AI could evolve to the point that humans could not control it, with Hawking theorizing that this could "spell the end of the human race".

One proposal to deal with this is to ensure that the first generally intelligent AI is 'Friendly AI', and will then be able to control subsequently developed AIs. Some question whether this kind of check could really remain in place.

Leading AI researcher Rodney Brooks writes, "I think it is a mistake to be worrying about us developing malevolent AI anytime in the next few hundred years. I think the worry stems from a fundamental error in not distinguishing the difference between the very real recent advances in a particular aspect of AI, and the enormity and complexity of building sentient volitional intelligence."

Machine Consciousness, Sentience and Mind

If an AI system replicates all key aspects of human intelligence, will that system also be sentient – will it have a mind which has conscious experiences? This question is closely related to the philosophical problem as to the nature of human consciousness, generally referred to as the hard problem of consciousness.

Consciousness

Computationalism and Functionalism

Computationalism is the position in the philosophy of mind that the human mind or the human brain (or both) is an information processing system and that thinking is a form of computing. Computationalism argues that the relationship between mind and body is similar or identical to the relationship between software and hardware and thus may be a solution to the mind-body problem. This philosophical position was inspired by the work of AI researchers and cognitive scientists in the 1960s and was originally proposed by philosophers Jerry Fodor and Hilary Putnam.

Strong AI Hypothesis

The philosophical position that John Searle has named "strong AI" states: "The appropriately programmed computer with the right inputs and outputs would thereby have a mind in exactly the same sense human beings have minds." Searle counters this assertion with his Chinese room argument, which asks us to look *inside* the computer and try to find where the "mind" might be.

Robot Rights

Mary Shelley's *Frankenstein* considers a key issue in the ethics of artificial intelligence: if a machine can be created that has intelligence, could it also *feel*? If it can feel, does it have the same rights as a human? The idea also appears in modern science fiction, such as the film *A.I.: Artificial Intelligence*, in which humanoid machines have the ability to feel emotions. This issue, now known as "robot rights", is currently being considered by, for example, California's Institute for the Future, although many critics believe that the discussion is premature. Some critics of transhumanism argue that any hypothetical robot rights would lie on a spectrum with animal rights and human rights. The subject is profoundly discussed in the 2010 documentary film *Plug & Pray*.

Superintelligence

Are there limits to how intelligent machines – or human-machine hybrids – can be? A superintelligence, hyperintelligence, or superhuman intelligence is a hypothetical agent that would possess intelligence far surpassing that of the brightest and most gifted human mind. "Superintelligence" may also refer to the form or degree of intelligence possessed by such an agent.

Technological Singularity

If research into Strong AI produced sufficiently intelligent software, it might be able to reprogram and improve itself. The improved software would be even better at improving itself, leading to recursive self-improvement. The new intelligence could thus increase exponentially and dramatically surpass humans. Science fiction writer Vernor Vinge named this scenario "singularity". Technological singularity is when accelerating progress in technologies will cause a runaway effect wherein artificial intelligence will exceed human intellectual capacity and control, thus radically changing or even ending civilization. Because the capabilities of such an intelligence may be impossible to comprehend, the technological singularity is an occurrence beyond which events are unpredictable or even unfathomable.

Ray Kurzweil has used Moore's law (which describes the relentless exponential improvement in digital technology) to calculate that desktop computers will have the same processing power as human brains by the year 2029, and predicts that the singularity will occur in 2045.

Transhumanism

You awake one morning to find your brain has another lobe functioning. Invisible, this auxiliary lobe answers your questions with information beyond the realm of your own memory, suggests plausible courses of action, and asks questions that help bring out relevant facts. You quickly come to rely on the new lobe so much that you stop wondering how it works. You just use it. This is the dream of artificial intelligence.

— Byte, April 1985

Robot designer Hans Moravec, cyberneticist Kevin Warwick and inventor Ray Kurzweil have predicted that humans and machines will merge in the future into cyborgs that are more capable and powerful than either. This idea, called transhumanism, which has roots in Aldous Huxley and Robert Ettinger, has been illustrated in fiction as well, for example in the manga *Ghost in the Shell* and the science-fiction series *Dune*.

In the 1980s artist Hajime Sorayama's Sexy Robots series were painted and published in Japan depicting the actual organic human form with lifelike muscular metallic skins and later "the Gynoids" book followed that was used by or influenced movie makers including George Lucas and other creatives. Sorayama never considered these organic robots to be real part of nature but always unnatural product of the human mind, a fantasy existing in the mind even when realized in actual form.

Edward Fredkin argues that "artificial intelligence is the next stage in evolution", an idea first proposed by Samuel Butler's "Darwin among the Machines" (1863), and expanded upon by George Dyson in his book of the same name in 1998.

In Fiction

Thought-capable artificial beings have appeared as storytelling devices since antiquity.

The implications of a constructed machine exhibiting artificial intelligence have been a persistent theme in science fiction since the twentieth century. Early stories typically revolved around intelligent robots. The word "robot" itself was coined by Karel Čapek in his 1921 play *R.U.R.*, the title standing for "Rossum's Universal Robots". Later, the SF writer Isaac Asimov developed the Three Laws of Robotics which he subsequently explored in a long series of robot stories. Asimov's laws are often brought up during layman discussions of machine ethics; while almost all artificial intelligence researchers are familiar with Asimov's laws through popular culture, they generally consider the laws useless for many reasons, one of which is their ambiguity.

The novel *Do Androids Dream of Electric Sheep?*, by Philip K. Dick, tells a science fiction story about Androids and humans clashing in a futuristic world. Elements of artificial intelligence include the empathy box, mood organ, and the androids themselves. Throughout the novel, Dick portrays the idea that human subjectivity is altered by technology created with artificial intelligence.

Nowadays AI is firmly rooted in popular culture; intelligent robots appear in innumerable works. HAL, the murderous computer in charge of the spaceship in *2001: A Space Odyssey* (1968), is an example of the common "robotic rampage" archetype in science fiction movies. *The Terminator* (1984) and *The Matrix* (1999) provide additional widely familiar examples. In contrast, the rare loyal robots such as Gort from *The Day the Earth Stood Still* (1951) and Bishop from *Aliens* (1986) are less prominent in popular culture.

Typical AI Problems

While studying the typical range of tasks that we might expect an "intelligent entity" to perform, we need to consider both "common-place" tasks as well as expert tasks.

Examples of common-place tasks include

- *Recognizing* people, objects.

    – Communicating (through *natural language*).

    – *Navigating* around obstacles on the streets

These tasks are done matter of factly and routinely by people and some other animals.

Expert tasks include:

- Medical diagnosis.
- Mathematical problem solving
- Playing games like chess

These tasks cannot be done by all people, and can only be performed by skilled specialists.

Now, which of these tasks are easy and which ones are hard? Clearly tasks of the first type are easy for humans to perform, and almost all are able to master them. The second range of tasks requires skill development and/or intelligence and only some specialists can perform them well. However, when we look at what computer systems have been able to achieve to date, we see that their achievements include performing sophisticated tasks like medical diagnosis, performing symbolic integration, proving theorems and playing chess.

On the other hand it has proved to be very hard to make computer systems perform many routine tasks that all humans and a lot of animals can do. Examples of such tasks include navigating our way without running into things, catching prey and avoiding predators. Humans and animals are also capable of interpreting complex sensory information. We are able to recognize objects and people from the visual image that we receive. We are also able to perform complex social functions.

Intelligent Behaviour

This discussion brings us back to the question of what constitutes intelligent behaviour. Some of these tasks and applications are:

- Perception involving image recognition and computer vision
- Reasoning
- Learning
- Understanding language involving natural language processing, speech processing
- Solving problems
- Robotics

Practical Impact of AI

AI components are embedded in numerous devices e.g. in copy machines for automatic correction of operation for copy quality improvement. AI systems are in everyday use for identifying credit card fraud, for advising doctors, for recognizing speech and in helping complex planning tasks.

Then there are intelligent tutoring systems that provide students with personalized attention Thus AI has increased understanding of the nature of intelligence and found many applications. It has helped in the understanding of human reasoning, and of the nature of intelligence. It has also helped us understand the complexity of modeling human reasoning.

Types of Artificial Intelligence

Applied AI: aims to produce commercially viable "smart" systems such as, for example, a security system that is able to recognise the faces of people who are permitted to enter a particular building. Applied AI has already enjoyed considerable success.

Cognitive AI: computers are used to test theories about how the human mind works--for example, theories about how we recognise faces and other objects, or about how we solve abstract problems.

Artificial General Intelligence

Artificial general intelligence (AGI) is the intelligence of a machine that could successfully perform any intellectual task that a human being can. It is a primary goal of some artificial intelligence research and a common topic in science fiction and futurism. Artificial general intelligence is also referred to as "strong AI", "full AI" or as the ability of a machine to perform "general intelligent action".

Some references emphasize a distinction between strong AI and "applied AI" (also called "narrow AI" or "weak AI"): the use of software to study or accomplish specific problem solving or reasoning tasks. Weak AI, in contrast to strong AI, does not attempt to perform the full range of human cognitive abilities.

Requirements

Many different definitions of intelligence have been proposed (such as being able to pass the Turing test) but to date, there is no definition that satisfies everyone. However, there *is* wide agreement among artificial intelligence researchers that intelligence is required to do the following:

- reason, use strategy, solve puzzles, and make judgments under uncertainty;

- represent knowledge, including commonsense knowledge;

- plan;

- learn;

- communicate in natural language;

- and integrate all these skills towards common goals.

Other important capabilities include the ability to sense (e.g. see) and the ability to act (e.g. move and manipulate objects) in the world where intelligent behaviour is to be observed. This would include an ability to detect and respond to hazard. Many interdisciplinary approaches to intelligence (e.g. cognitive science, computational intelligence and decision making) tend to emphasise the need to consider additional traits such as imagination (taken as the ability to form mental images and concepts that were not programmed in) and autonomy. Computer based systems that exhibit many of these capabilities do exist (e.g. computational creativity, automated reasoning, decision support system, robot, evolutionary computation, intelligent agent), but not yet at human levels.

Tests for Confirming Operational AGI

Scientists have varying ideas of what kinds of tests a human-level intelligent machine needs to pass in order to be considered an operational example of artificial general intelligence. A few of these scientists include the late Alan Turing, Steve Wozniak, Ben Goertzel, and Nils Nilsson. A few of the tests they have proposed are:

The Turing Test (Turing)

In the Turing Test, a machine and a human both converse sight unseen with a second human, who must evaluate which of the two is the machine.

The Coffee Test (Wozniak)

A machine is given the task of going into an average American home and figuring out how to make coffee. It has to find the coffee machine, find the coffee, add water, find a mug, and brew the coffee by pushing the proper buttons.

The Robot College Student Test (Goertzel)

A machine is given the task of enrolling in a university, taking and passing the same classes that humans would, and obtaining a degree.

The Employment Test (Nilsson)

A machine is given the task of working an economically important job, and must perform as well or better than the level that humans perform at in the same job.

These are a few tests that cover a variety of qualities that a machine might need to have to be considered AGI, including the ability to reason and learn.

Problems Requiring AGI to Solve

The most difficult problems for computers to solve are informally known as "AI-complete" or "AI-hard", implying that the difficulty of these computational problems is equivalent to that of solving the central artificial intelligence problem—making computers as intelligent as people, or strong AI. To call a problem AI-complete reflects an attitude that it would not be solved by a simple specific algorithm.

AI-complete problems are hypothesised to include computer vision, natural language understanding, and dealing with unexpected circumstances while solving any real world problem.

Currently, AI-complete problems cannot be solved with modern computer technology alone, and also require human computation. This property can be useful, for instance to test for the presence of humans as with CAPTCHAs, and for computer security to circumvent brute-force attacks.

Mainstream AI research

History of Mainstream Research into Strong AI

Modern AI research began in the mid 1950s. The first generation of AI researchers were convinced that strong AI was possible and that it would exist in just a few decades. As AI pioneer Herbert A. Simon wrote in 1965: "machines will be capable, within twenty years, of doing any work a man can do." Their predictions were the inspiration for Stanley Kubrick and Arthur C. Clarke's character HAL 9000, who accurately embodied what AI researchers believed they could create by the year 2001. Of note is the fact that AI pioneer Marvin Minsky was a consultant on the project of making HAL 9000 as realistic as possible according to the consensus predictions of the time; Crevier quotes him as having said on the subject in 1967, "Within a generation...the problem of creating 'artificial intelligence' will substantially be solved,", although Minsky states that he was misquoted.

However, in the early 1970s, it became obvious that researchers had grossly underestimated the difficulty of the project. The agencies that funded AI became skeptical of strong AI and put researchers under increasing pressure to produce useful technology, or "applied AI". As the 1980s began, Japan's fifth generation computer project revived interest in strong AI, setting out a ten-year timeline that included strong AI goals like "carry on a casual conversation". In response to this and the success of expert systems, both industry and government pumped money back into the field. However, the market for AI spectacularly collapsed in the late 1980s and the goals of the fifth generation computer project were never fulfilled. For the second time in 20 years, AI researchers who had predicted the imminent arrival of strong AI had been shown to be fundamentally mistaken about what they could accomplish. By the 1990s, AI researchers had gained a reputation for making promises they could not keep. AI researchers became reluctant to make any kind of prediction at all and avoid any mention of "human level" artificial intelligence, for fear of being labeled a "wild-eyed dreamer."

Current Mainstream AI Research

In the 1990s and early 21st century, mainstream AI has achieved a far higher degree of commercial success and academic respectability by focusing on specific sub-problems where they can produce verifiable results and commercial applications, such as neural networks, computer vision or data mining. These "applied AI" applications are now used extensively throughout the technology industry and research in this vein is very heavily funded in both academia and industry.

Most mainstream AI researchers hope that strong AI can be developed by combining the programs that solve various sub-problems using an integrated agent architecture, cognitive architecture or subsumption architecture. Hans Moravec wrote in 1988:

"I am confident that this bottom-up route to artificial intelligence will one day meet the traditional top-down route more than half way, ready to provide the real world competence and the common-sense knowledge that has been so frustratingly elusive in reasoning programs. Fully intelligent machines will result when the metaphorical golden spike is driven uniting the two efforts."

However, much contention has existed in AI research, even with regards to the fundamental philosophies informing this field; for example, Stevan Harnad from Princeton stated in the conclusion of his 1990 paper on the Symbol Grounding Hypothesis that:

"The expectation has often been voiced that "top-down" (symbolic) approaches to modeling cognition will somehow meet "bottom-up" (sensory) approaches somewhere in between. If the grounding considerations in this paper are valid, then this expectation is hopelessly modular and there is really only one viable route from sense to symbols: from the ground up. A free-floating symbolic level like the software level of a computer will never be reached by this route (or vice versa) -- nor is it clear why we should even try to reach such a level, since it looks as if getting there would just amount to uprooting our symbols from their intrinsic meanings (thereby merely reducing ourselves to the functional equivalent of a programmable computer)."

Artificial General Intelligence Research

Artificial general intelligence (AGI) describes research that aims to create machines capable of general intelligent action. The term was introduced by Mark Gubrud in 1997 in a discussion of the implications of fully automated military production and operations. The research objective is much older, for example Doug Lenat's Cyc project (that began in 1984), and Allen Newell's Soar project are regarded as within the scope of AGI. AGI research activity in 2006 was described by Pei Wang and Ben Goertzel as "producing publications and preliminary results". As yet, most AI researchers have devoted little attention to AGI, with some claiming that intelligence is too complex to be completely replicated in the near term. However, a small number of computer scientists are active in AGI research, and many of this group are contributing to a series of AGI conferences. The research is extremely diverse and often pioneering in nature. In the introduction to his book, Goertzel says that estimates of the time needed before a truly flexible AGI is built vary from 10 years to over a century, but the consensus in the AGI research community seems to be that the timeline discussed by Ray Kurzweil in *The Singularity is Near* (i.e. between 2015 and 2045) is plausible. Most mainstream AI researchers doubt that progress will be this rapid. Organizations actively pursuing AGI include the Machine Intelligence Research Institute, the OpenCog Foundation, the Swiss AI lab IDSIA, Numenta and the associated Redwood Neuroscience Institute.

Processing Power Needed to Simulate a Brain

Whole Brain Emulation

A popular approach discussed to achieving general intelligent action is whole brain emulation. A low-level brain model is built by scanning and mapping a biological brain in detail and copying its state into a computer system or another computational device. The computer runs a simulation model so faithful to the original that it will behave in essentially the same way as the original brain, or for all practical purposes, indistinguishably. Whole brain emulation is discussed in computational neuroscience and neuroinformatics, in the context of brain simulation for medical research purposes. It is discussed in artificial intelligence research as an approach to strong AI. Neuroimaging technologies that could deliver the necessary detailed understanding are improving rapidly, and futurist Ray Kurzweil in the book *The Singularity Is Near* predicts that a map of sufficient quality will become available on a similar timescale to the required computing power.

Early Estimates

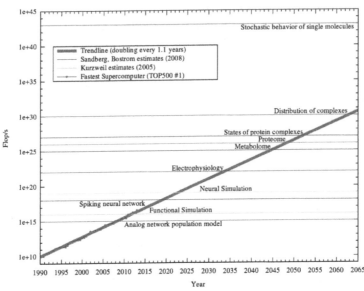

Estimates of how much processing power is needed to emulate a human brain at various levels (from Ray Kurzweil, and Anders Sandberg and Nick Bostrom), along with the fastest supercomputer from TOP500 mapped by year. Note the logarithmic scale and exponential trendline, which assumes the computational capacity doubles every 1.1 years. Kurzweil believes that mind uploading will be possible at neural simulation, while the Sandberg, Bostrom report is less certain about where consciousness arises.

For low-level brain simulation, an extremely powerful computer would be required. The human brain has a huge number of synapses. Each of the 10^{11} (one hundred billion) neurons has on average 7,000 synaptic connections to other neurons. It has been estimated that the brain of a three-year-old child has about 10^{15} synapses (1 quadrillion). This number declines with age, stabilizing by adulthood. Estimates vary for an adult, ranging from 10^{14} to 5×10^{14} synapses (100 to 500 trillion). An estimate of the brain's processing power, based on a simple switch model for neuron activity, is around 10^{14} (100 trillion) synaptic updates per second (SUPS). In 1997 Kurzweil looked at various estimates for the hardware required to equal the human brain and adopted a figure of 10^{16} computations per second (cps). (For comparison, if a "computation" was equivalent to one "Floating Point Operation" - a measure used to rate current supercomputers - then 10^{16} "computations" would be equivalent to 10 PetaFLOPS, achieved in 2011). He uses this figure to predict the necessary hardware will be available sometime between 2015 and 2025, if the current exponential growth in computer power continues.

Modelling the Neurons in More Detail

The artificial neuron model assumed by Kurzweil and used in many current artificial neural network implementations is simple compared with biological neurons. A brain simulation would likely have to capture the detailed cellular behaviour of biological neurons, presently only understood in the broadest of outlines. The overhead introduced by full modeling of the biological, chemical, and physical details of neural behaviour (especially on a molecular scale) would require computational powers several orders of magnitude larger than Kurzweil's estimate. In addition the estimates do not account for Glial cells which are at least as numerous as neurons, may outnumber neurons by as much as 10:1, and are now known to play a role in cognitive processes.

Current Research

There are some research projects that are investigating brain simulation using more sophisticated neural models, implemented on conventional computing architectures. The Artificial Intelligence System project implemented non-real time simulations of a "brain" (with 10^{11} neurons) in 2005. It took 50 days on a cluster of 27 processors to simulate 1 second of a model. The Blue Brain project used one of the fastest supercomputer architectures in the world, IBM's Blue Gene platform, to create a real time simulation of a single rat neocortical column consisting of approximately 10,000 neurons and 10^8 synapses in 2006. A longer term goal is to build a detailed, functional simulation of the physiological processes in the human brain: "It is not impossible to build a human brain and we can do it in 10 years," Henry Markram, director of the Blue Brain Project said in 2009 at the TED conference in Oxford. There have also been controversial claims to have simulated a cat brain. Neuro-silicon interfaces have been proposed as an alternative implementation strategy that may scale better.

Hans Moravec addressed the above arguments ("brains are more complicated", "neurons have to be modeled in more detail") in his 1997 paper "When will computer hardware match the human brain?". He measured the ability of existing software to simulate the functionality of neural tissue, specifically the retina. His results do not depend on the number of glial cells, nor on what kinds of processing neurons perform where.

Complications and Criticisms of AI Approaches based on Simulation

A fundamental criticism of the simulated brain approach derives from embodied cognition where human embodiment is taken as an essential aspect of human intelligence. Many researchers believe that embodiment is necessary to ground meaning. If this view is correct, any fully functional brain model will need to encompass more than just the neurons (i.e., a robotic body). Goertzel proposes virtual embodiment (like Second Life), but it is not yet known whether this would be sufficient.

Desktop computers using microprocessors capable of more than 10^9 cps have been available since 2005. According to the brain power estimates used by Kurzweil (and Moravec), this computer should be capable of supporting a simulation of a bee brain, but despite some interest no such simulation exists. There are at least three reasons for this:

- Firstly, the neuron model seems to be oversimplified.

- Secondly, there is insufficient understanding of higher cognitive processes to establish accurately what the brain's neural activity, observed using techniques such as functional magnetic resonance imaging, correlates with.

- Thirdly, even if our understanding of cognition advances sufficiently, early simulation programs are likely to be very inefficient and will, therefore, need considerably more hardware.

- Fourthly, the brain of an organism, while critical, may not be an appropriate boundary for a cognitive model. To simulate a bee brain, it may be necessary to simulate the body, and the environment. The Extended Mind thesis formalizes the philosophical concept, and research into cephalopods has demonstrated clear examples of a decentralized system.

In addition, the scale of the human brain is not currently well-constrained. One estimate puts the human brain at about 100 billion neurons and 100 trillion synapses. Another estimate is 86 billion neurons of which 16.3 billion are in the cerebral cortex and 69 billion in the cerebellum. Glial cell synapses are currently unquantified but are known to be extremely numerous.

Artificial Consciousness Research

Although the role of consciousness in strong AI/AGI is debatable, many AGI researchers regard research that investigates possibilities for implementing consciousness as vital. In an early effort Igor Aleksander argued that the principles for creating a conscious machine already existed but that it would take forty years to train such a machine to understand language.

Relationship to "Strong AI"

In 1980, philosopher John Searle coined the term "strong AI" as part of his Chinese room argument. He wanted to distinguish between two different hypotheses about artificial intelligence:

- An artificial intelligence system can *think* and have a *mind*. (The word "mind" has a specific meaning for philosophers, as used in "the mind body problem" or "the philosophy of mind".)

- An artificial intelligence system can (only) *act like* it thinks and has a mind.

The first one is called "the *strong* AI hypothesis" and the second is "the *weak* AI hypothesis" because the first one makes the *stronger* statement: it assumes something special has happened to the machine that goes beyond all its abilities that we can test. Searle referred to the "strong AI hypothesis" as "strong AI". This usage is also common in academic AI research and textbooks.

The weak AI hypothesis is equivalent to the hypothesis that artificial general intelligence is possible. According to Russell and Norvig, "Most AI researchers take the weak AI hypothesis for granted, and don't care about the strong AI hypothesis."

In contrast to Searle, Kurzweil uses the term "strong AI" to describe any artificial intelligence system that acts like it has a mind, regardless of whether a philosopher would be able to determine if it *actually* has a mind or not.

Possible Explanations for the Slow Progress of AI Research

Since the launch of AI research in 1956, the growth of this field has slowed down over time and has stalled the aims of creating machines skilled with intelligent action at the human level. A possible explanation for this delay is that computers lack a sufficient scope of memory or processing power.

In addition, the level of complexity that connects to the process of AI research may also limit the progress of AI research.

While most AI researchers believe that strong AI can be achieved in the future, there are some individuals like Hubert Dreyfus and Roger Penrose that deny the possibility of achieving AI. John McCarthy was one of various computer scientists who believe human-level AI will be accomplished, but a date cannot accurately be predicted.

Conceptual limitations are another possible reason for the slowness in AI research. AI researchers may need to modify the conceptual framework of their discipline in order to provide a stronger base and contribution to the quest of achieving strong AI. As William Clocksin wrote in 2003: "the framework starts from Weizenbaum's observation that intelligence manifests itself only relative to specific social and cultural contexts".

Furthermore, AI researchers have been able to create computers that can perform jobs that are complicated for people to do, but conversely they have struggled to develop a computer that is capable of carrying out tasks that are simple for humans to do. A problem that is described by David Gelernter is that some people assume that thinking and reasoning are equivalent. However, the idea of whether thoughts and the creator of those thoughts are isolated individually has intrigued AI researchers.

The problems that have been encountered in AI research over the past decades have further impeded the progress of AI. The failed predictions that have been promised by AI researchers and the lack of a complete understanding of human behaviors have helped diminish the primary idea of human-level AI. Although the progress of AI research has brought both improvement and disappointment, most investigators have established optimism about potentially achieving the goal of AI in the 21st century.

Other possible reasons have been proposed for the lengthy research in the progress of strong AI. The intricacy of scientific problems and the need to fully understand the human brain through psychology and neurophysiology have limited many researchers from emulating the function of the human brain into a computer hardware. Many researchers tend to underestimate any doubt that is involved with future predictions of AI, but without taking those issues seriously can people then overlook solutions to problematic questions.

Clocksin says that a conceptual limitation that may impede the progress of AI research is that people may be using the wrong techniques for computer programs and implementation of equipment. When AI researchers first began to aim for the goal of artificial intelligence, a main interest was human reasoning. Researchers hoped to establish computational models of human knowledge through reasoning and to find out how to design a computer with a specific cognitive task.

The practice of abstraction, which people tend to redefine when working with a particular context in research, provides researchers with a concentration on just a few concepts. The most productive use of abstraction in AI research comes from planning and problem solving. Although the aim is to increase the speed of a computation, the role of abstraction has posed questions about the involvement of abstraction operators.

A possible reason for the slowness in AI relates to the acknowledgement by many AI researchers that heuristics is a section that contains a significant breach between computer performance and human performance. The specific functions that are programmed to a computer may be able to account for many of the requirements that allow it to match human intelligence. These explanations are not necessarily guaranteed to be the fundamental causes for the delay in achieving strong AI, but they are widely agreed by numerous researchers.

There have been many AI researchers that debate over the idea whether machines should be created with emotions. There are no emotions in typical models of AI and some researchers say programming emotions into machines allows them to have a mind of their own. Emotion sums up the experiences of humans because it allows them to remember those experiences. David Gelernter writes, "No computer will be creative unless it can simulate all the nuances of human emotion." This concern about emotion has posed problems for AI researchers and it connects to the concept of strong AI as its research progresses into the future.

Consciousness

There are other aspects of the human mind besides intelligence that are relevant to the concept of strong AI which play a major role in science fiction and the ethics of artificial intelligence:

- consciousness: To have subjective experience and thought.

- self-awareness: To be aware of oneself as a separate individual, especially to be aware of one's own thoughts.

- sentience: The ability to "feel" perceptions or emotions subjectively.

- sapience: The capacity for wisdom.

These traits have a moral dimension, because a machine with this form of strong AI may have legal rights, analogous to the rights of non-human animals. Also, Bill Joy, among others, argues a machine with these traits may be a threat to human life or dignity. It remains to be shown whether any of these traits are necessary for strong AI. The role of consciousness is not clear, and currently there is no agreed test for its presence. If a machine is built with a device that simulates the neural correlates of consciousness, would it automatically have self-awareness? It is also possible that some of these properties, such as sentience, naturally emerge from a fully intelligent machine, or that it becomes natural to *ascribe* these properties to machines once they begin to act in a way that is clearly intelligent. For example, intelligent action may be sufficient for sentience, rather than the other way around.

In science fiction, AGI is associated with traits such as consciousness, sentience, sapience, and self-awareness observed in living beings. However, according to philosopher John Searle, it is an open question whether general intelligence is sufficient for consciousness, even a digital brain simulation. "Strong AI" (as defined above by Ray Kurzweil) should not be confused with Searle's "'strong AI hypothesis". The strong AI hypothesis is the claim that a computer which behaves as intelligently as a person must also necessarily have a mind and consciousness. AGI refers only to the amount of intelligence that the machine displays, with or without a mind.

Controversies and Dangers

Feasibility

Opinions vary both on *whether* and *when* artificial general intelligence will arrive. At one extreme, AI pioneer Herbert A. Simon wrote in 1965: "machines will be capable, within twenty years, of doing any work a man can do"; obviously this prediction failed to come true. Microsoft co-founder Paul Allen believes that such intelligence is unlikely this century because it would require "unforeseeable and fundamentally unpredictable breakthroughs" and a "scientifically deep understanding of cognition". Writing in The Guardian, roboticist Alan Winfield claimed the gulf between modern computing and human-level artificial intelligence is as wide as the gulf between current space flight and practical faster than light spaceflight. Optimism that AGI is feasible waxes and wanes, and may have seen a resurgence in the 2010s: around 2015, computer scientist Richard Sutton averaged together some recent polls of artificial intelligence experts and estimated a 25% chance that AGI will arrive before 2030, but a 10% chance that it will never arrive at all.

Risk of Human Extinction

The creation of artificial general intelligence may have repercussions so great and so complex that it may not be possible to forecast what will come afterwards. Thus the event in the hypothetical future of achieving strong AI is called the technological singularity, because theoretically one cannot see past it. But this has not stopped philosophers and researchers from guessing what the smart computers or robots of the future may do, including forming a utopia by being our friends or overwhelming us in an AI takeover. The latter potentiality is particularly disturbing as it poses an existential risk for mankind.

Self-replicating Machines

Smart computers or robots would be able to produce copies of themselves. They would be self-replicating machines. A growing population of intelligent robots could conceivably outcompete inferior humans in job markets, in business, in science, in politics (pursuing robot rights), and technologically, sociologically (by acting as one), and militarily.

Emergent Superintelligence

If research into strong AI produced sufficiently intelligent software, it would be able to reprogram and improve itself – a feature called "recursive self-improvement". It would then be even better at improving itself, and would probably continue doing so in a rapidly increasing cycle, leading to an intelligence explosion and the emergence of superintelligence. Such an intelligence would not have the limitations of human intellect, and might be able to invent or discover almost anything.

Hyper-intelligent software might not necessarily decide to support the continued existence of mankind, and might be extremely difficult to stop. This topic has also recently begun to be discussed in academic publications as a real source of risks to civilization, humans, and planet Earth.

One proposal to deal with this is to make sure that the first generally intelligent AI is friendly AI, that would then endeavor to ensure that subsequently developed AIs were also nice to us. But, friendly AI is harder to create than plain AGI, and therefore it is likely, in a race between the two, that non-friendly AI would be developed first. Also, there is no guarantee that friendly AI would remain friendly, or that its progeny would also all be good.

Weak AI

Weak AI (also known as narrow AI) is non-sentient artificial intelligence that is focused on one narrow task. Weak AI is defined in contrast to either strong AI (a machine with consciousness, sentience and mind) or artificial general intelligence (a machine with the ability to apply intelligence to any problem, rather than just one specific problem). All currently existing systems considered artificial intelligence of any sort are weak AI at most.

Siri is a good example of narrow intelligence. Siri operates within a limited pre-defined range, there is no genuine intelligence, no self-awareness, no life despite being a sophisticated example of weak AI. In Forbes (2011), Ted Greenwald wrote: "The iPhone/Siri marriage represents the arrival of hybrid AI, combining several narrow AI techniques plus access to massive data in the cloud." AI researcher Ben Goertzel, on his blog in 2010, stated Siri was "VERY narrow and brittle" evidenced by annoying results if you ask questions outside the limits of the application.

Some commentators think weak AI could be dangerous. In 2013 George Dvorsky stated via io9: "Narrow AI could knock out our electric grid, damage nuclear power plants, cause a global-scale economic collapse, misdirect autonomous vehicles and robots..." The Stanford Center for Internet and Society, in the following quote, contrasts strong AI with weak AI regarding the growth of narrow AI presenting "real issues".

Weak or "narrow" AI, in contrast, is a present-day reality. Software controls many facets of daily life and, in some cases, this control presents real issues. One example is the May 2010 "flash crash" that caused a temporary but enormous dip in the market.

> *—Ryan Calo, Center for Internet and Society, Stanford Law School, 30 August 2011.*

The following two excerpts from Singularity Hub summarise weak-narrow AI:

When you call the bank and talk to an automated voice you are probably talking to an AI...just a very annoying one. Our world is full of these limited AI programs which we classify as "weak" or "narrow" or "applied". These programs are far from the sentient, love-seeking, angst-ridden artificial intelligences we see in science fiction, but that's temporary. All these narrow AIs are like the amino acids in the primordial ooze of the Earth.

We're slowly building a library of narrow AI talents that are becoming more impressive. Speech recognition and processing allows computers to convert sounds to text with greater accuracy. Google is using AI to caption millions of videos on YouTube. Likewise, computer vision is improving so that programs like Vitamin d Video can recognize objects, classify them, and understand how they move. Narrow AI isn't just getting better at processing its environment it's also understanding the difference between what a human says and what a human wants.

> *—Aaron Saenz, Singularity Hub, 10 August 2010.*

Limits of AI

Today's successful AI systems operate in well-defined domains and employ narrow, specialized knowledge. Common sense knowledge is needed to function in complex, open-ended worlds. Such a system also needs to understand unconstrained natural language. However these capabilities are not yet fully present in today's intelligent systems.

What can AI systems do?

Today's AI systems have been able to achieve limited success in some of these tasks.

- In Computer vision, the systems are capable of face recognition

- In Robotics, we have been able to make vehicles that are mostly autonomous.

- In Natural language processing, we have systems that are capable of simple machine translation.

- Today's Expert systems can carry out medical diagnosis in a narrow domain

- Speech understanding systems are capable of recognizing several thousand words continuous speech

- Planning and scheduling systems had been employed in scheduling experiments with the Hubble Telescope.

- The Learning systems are capable of doing text categorization into about a 1000 topics

- In Games, AI systems can play at the Grand Master level in chess (world champion), checkers, etc.

What can AI systems NOT do yet?

Understand natural language robustly (e.g., read and understand articles in a newspaper)

- Surf the web

- Interpret an arbitrary visual scene

- Learn a natural language

- Construct plans in dynamic real-time domains

- Exhibit true autonomy and intelligence

AI History

Intellectual roots of AI date back to the early studies of the nature of knowledge and reasoning. The dream of making a computer imitate humans also has a very early history.

The concept of intelligent machines is found in Greek mythology. There is a story in the 8th century A.D about Pygmalion Olio, the legendary king of Cyprus. He fell in love with an ivory statue he made to represent his ideal woman. The king prayed to the goddess Aphrodite, and the goddess miraculously brought the statue to life. Other myths involve human-like artifacts. As a present from Zeus to Europa, Hephaestus created Talos, a huge robot. Talos was made of bronze and his duty was to patrol the beaches of Crete.

Aristotle (384-322 BC) developed an informal system of syllogistic logic, which is the basis of the first formal deductive reasoning system.

Early in the 17th century, Descartes proposed that bodies of animals are nothing more than complex machines.

Pascal in 1642 made the first mechanical digital calculating machine.

In the 19th century, George Boole developed a binary algebra representing (some) "laws of thought."

Charles Babbage & Ada Byron worked on programmable mechanical calculating machines.

In the late 19th century and early 20th century, mathematical philosophers like Gottlob Frege, Bertram Russell, Alfred North Whitehead, and Kurt Gödel built on Boole's initial logic concepts to develop mathematical representations of logic problems.

The advent of electronic computers provided a revolutionary advance in the ability to study intelligence.

In 1943 McCulloch & Pitts developed a Boolean circuit model of brain. They wrote the paper "A Logical Calculus of Ideas Immanent in Nervous Activity", which explained how it is possible for neural networks to compute.

Marvin Minsky and Dean Edmonds built the SNARC in 1951, which is the first randomly wired neural network learning machine (SNARC stands for Stochastic Neural-Analog Reinforcement Computer).It was a neural network computer that used 3000 vacuum tubes and a network with 40 neurons.

In 1950 Turing wrote an article on "Computing Machinery and Intelligence" which articulated a complete vision of AI. Turing's paper talked of many things, of solving problems by searching through the space of possible solutions, guided by heuristics. He illustrated his ideas on machine intelligence by reference to chess. He even propounded the possibility of letting the machine alter its own instructions so that machines can learn from experience.

In 1956 a famous conference took place in Dartmouth. The conference brought together the founding fathers of artificial intelligence for the first time. In this meeting the term "Artificial Intelligence" was adopted.

Between 1952 and 1956, Samuel had developed several programs for playing checkers. In 1956, Newell & Simon's Logic Theorist was published. It is considered by many to be the first AI program. In 1959, Gelernter developed a Geometry Engine. In 1961 James Slagle (PhD dissertation, MIT) wrote a symbolic integration program, SAINT. It was written in LISP and solved calculus problems at the college freshman level. In 1963, Thomas Evan's program Analogy was developed which could solve IQ test type analogy problems.

In 1963, Edward A. Feigenbaum & Julian Feldman published Computers and Thought, the first collection of articles about artificial intelligence.

In 1965, J. Allen Robinson invented a mechanical proof procedure, the Resolution Method, which allowed programs to work efficiently with formal logic as a representation language. In 1967, the Dendral program (Feigenbaum, Lederberg, Buchanan, Sutherland at Stanford) was demonstrated which could interpret mass spectra on organic chemical compounds. This was the first successful knowledge-based program for scientific reasoning. In 1969 the SRI robot, Shakey, demonstrated combining locomotion, perception and problem solving.

The years from 1969 to 1979 marked the early development of knowledge-based systems.

In 1974: MYCIN demonstrated the power of rule-based systems for knowledge representation and inference in medical diagnosis and therapy. Knowledge representation schemes were developed. These included frames developed by Minski. Logic based languages like Prolog and Planner were developed.

In the 1980s, Lisp Machines developed and marketed.

Around 1985, neural networks return to popularity.

In 1988, there was a resurgence of probabilistic and decision-theoretic methods.

The early AI systems used general systems, little knowledge. AI researchers realized that specialized knowledge is required for rich tasks to focus reasoning.

The 1990's saw major advances in all areas of AI including the following:

- machine learning, data mining
- intelligent tutoring,
- case-based reasoning,
- multi-agent planning, scheduling,
- uncertain reasoning,
- natural language understanding and translation,
- vision, virtual reality, games, and other topics.

Rod Brooks' COG Project at MIT, with numerous collaborators, made significant progress in building a humanoid robot.

The first official Robo-Cup soccer match featuring table-top matches with 40 teams of interacting robots was held in 1997.

In the late 90s, Web crawlers and other AI-based information extraction programs become essential in widespread use of the world-wide-web.

Interactive robot pets ("smart toys") become commercially available, realizing the vision of the 18th century novelty toy makers.

In 2000, the Nomad robot explores remote regions of Antarctica looking for meteorite samples.

We will now look at a few famous AI system that has been developed over the years.

1. ALVINN:

Autonomous Land Vehicle In a Neural Network

In 1989, Dean Pomerleau at CMU created ALVINN. This is a system which learns to control vehicles by watching a person drive. It contains a neural network whose input is a 30x32 unit two dimensional camera image. The output layer is a representation of the direction the vehicle should travel.

The system drove a car from the East Coast of USA to the west coast, a total of about 2850 miles. Out of this about 50 miles were driven by a human, and the rest solely by the system.

2. Deep Blue

In 1997, the Deep Blue chess program created by IBM, beat the current world chess champion, Gary Kasparov.

3. Machine translation

A system capable of translations between people speaking different languages will be a remarkable achievement of enormous economic and cultural benefit. Machine translation is one of the important fields of endeavour in AI. While some translating systems have been developed, there is a lot of scope for improvement in translation quality.

4. Autonomous agents

In space exploration, robotic space probes autonomously monitor their surroundings, make decisions and act to achieve their goals.

NASA's Mars rovers successfully completed their primary three-month missions in April, 2004. The Spirit rover had been exploring a range of Martian hills that took two months to reach. It is finding curiously eroded rocks that may be new pieces to the puzzle of the region's past. Spirit's twin, Opportunity, had been examining exposed rock layers inside a crater.

5. Internet agents

The explosive growth of the internet has also led to growing interest in internet agents to monitor users' tasks, seek needed information, and to learn which information is most useful.

References

- Hernandez-Orallo, J.; Dowe, D. L. (2010). "Measuring Universal Intelligence: Towards an Anytime Intelligence Test". Artificial Intelligence Journal. 174 (18): 1508–1539. doi:10.1016/j.artint.2010.09.006
- Hutter, M. (2012). "One Decade of Universal Artificial Intelligence". Theoretical Foundations of Artificial Gen-

eral Intelligence. Atlantis Thinking Machines. 4. doi:10.2991/978-94-91216-62-6_5. ISBN 978-94-91216-61-9

- Clocksin, William (Aug 2003), "Artificial intelligence and the future", Philosophical Transactions of the Royal Society A, 361 (1809): 1721–1748, doi:10.1098/rsta.2003.1232, PMID 12952683

- Lakoff, George; Núñez, Rafael E. (2000). Where Mathematics Comes From: How the Embodied Mind Brings Mathematics into Being. Basic Books. ISBN 0-465-03771-2

- McCarthy, John; Minsky, Marvin; Rochester, Nathan; Shannon, Claude (1955). "A Proposal for the Dartmouth Summer Research Project on Artificial Intelligence". Archived from the original on 26 August 2007. Retrieved 30 August 2007

- Feigenbaum, Edward A.; McCorduck, Pamela (1983), The Fifth Generation: Artificial Intelligence and Japan's Computer Challenge to the World, Michael Joseph, ISBN 0-7181-2401-4

- McCarthy, John; Hayes, P. J. (1969). "Some philosophical problems from the standpoint of artificial intelligence". Machine Intelligence. 4: 463–502. Archived from the original on 10 August 2007. Retrieved 30 August 2007

- Luger, George; Stubblefield, William (2004), Artificial Intelligence: Structures and Strategies for Complex Problem Solving (5th ed.), The Benjamin/Cummings Publishing Company, Inc., p. 720, ISBN 0-8053-4780-1

- Turing, Alan (October 1950), "Computing Machinery and Intelligence", Mind, LIX (236): 433–460, doi:10.1093/mind/LIX.236.433, ISSN 0026-4423, retrieved 2008-08-18

- Müller, Vincent C.; Bostrom, Nick (2014). "Future Progress in Artificial Intelligence: A Poll Among Experts" Yudkowsky, Eliezer (2006), Goertzel, Ben; Pennachin, Cassio, eds., Artificial General Intelligence (PDF), 49, Springer, pp. 585–612, ISBN 3-540-23733-X, doi:10.1146/annurev.psych.49.1.585, archived from the original (PDF) on 11 April 2009

- Holley, Peter (28 January 2015). "Bill Gates on dangers of artificial intelligence: 'I don't understand why some people are not concerned'". The Washington Post. ISSN 0190-8286. Retrieved 2015-10-30

Artificial Intelligence: Knowledge Representation and Logic

Knowledge representation and logic commutes information about the world to a machine in a format that can easily be understood by them. Examples of knowledge representation and logic include systems architecture, ontologies, frames and semantic nets. This section discusses the methods of information representation and logic in a critical manner providing key analysis to the subject matter.

Knowledge Representation and Reasoning

Knowledge representation and reasoning (KR) is the field of artificial intelligence (AI) dedicated to representing information about the world in a form that a computer system can utilize to solve complex tasks such as diagnosing a medical condition or having a dialog in a natural language. Knowledge representation incorporates findings from psychology about how humans solve problems and represent knowledge in order to design formalisms that will make complex systems easier to design and build. Knowledge representation and reasoning also incorporates findings from logic to automate various kinds of *reasoning*, such as the application of rules or the relations of sets and subsets.

Examples of knowledge representation formalisms include semantic nets, systems architecture, frames, rules, and ontologies. Examples of automated reasoning engines include inference engines, theorem provers, and classifiers.

History

The earliest work in computerized knowledge representation was focused on general problem solvers such as the General Problem Solver (GPS) system developed by Allen Newell and Herbert A. Simon in 1959. These systems featured data structures for planning and decomposition. The system would begin with a goal. It would then decompose that goal into sub-goals and then set out to construct strategies that could accomplish each subgoal.

In these early days of AI, general search algorithms such as A* were also developed. However, the amorphous problem definitions for systems such as GPS meant that they worked only for very constrained toy domains (e.g. the "blocks world"). In order to tackle non-toy problems, AI researchers such as Ed Feigenbaum and Frederick Hayes-Roth realized that it was necessary to focus systems on more constrained problems.

It was the failure of these efforts that led to the cognitive revolution in psychology and to the phase of AI focused on knowledge representation that resulted in expert systems in the 1970s and 80s, production systems, frame languages, etc. Rather than general problem solvers, AI changed its focus to expert systems that could match human competence on a specific task, such as medical diagnosis.

Expert systems gave us the terminology still in use today where AI systems are divided into a Knowledge Base with facts about the world and rules and an inference engine that applies the rules to the knowledge base in order to answer questions and solve problems. In these early systems the knowledge base tended to be a fairly flat structure, essentially assertions about the values of variables used by the rules.

In addition to expert systems, other researchers developed the concept of frame based languages in the mid 1980s. A frame is similar to an object class: It is an abstract description of a category describing things in the world, problems, and potential solutions. Frames were originally used on systems geared toward human interaction, e.g. understanding natural language and the social settings in which various default expectations such as ordering food in a restaurant narrow the search space and allow the system to choose appropriate responses to dynamic situations.

It wasn't long before the frame communities and the rule-based researchers realized that there was synergy between their approaches. Frames were good for representing the real world, described as classes, subclasses, slots (data values) with various constraints on possible values. Rules were good for representing and utilizing complex logic such as the process to make a medical diagnosis. Integrated systems were developed that combined Frames and Rules. One of the most powerful and well known was the 1983 Knowledge Engineering Environment (KEE) from Intellicorp. KEE had a complete rule engine with forward and backward chaining. It also had a complete frame based knowledge base with triggers, slots (data values), inheritance, and message passing. Although message passing originated in the object-oriented community rather than AI it was quickly embraced by AI researchers as well in environments such as KEE and in the operating systems for Lisp machines from Symbolics, Xerox, and Texas Instruments.

The integration of Frames, rules, and object-oriented programming was significantly driven by commercial ventures such as KEE and Symbolics spun off from various research projects. At the same time as this was occurring, there was another strain of research which was less commercially focused and was driven by mathematical logic and automated theorem proving. One of the most influential languages in this research was the KL-ONE language of the mid 80's. KL-ONE was a frame language that had a rigorous semantics, formal definitions for concepts such as an Is-A relation. KL-ONE and languages that were influenced by it such as Loom had an automated reasoning engine that was based on formal logic rather than on IF-THEN rules. This reasoner is called the classifier. A classifier can analyze a set of declarations and infer new assertions, for example, redefine a class to be a subclass or superclass of some other class that wasn't formally specified. In this way the classifier can function as an inference engine, deducing new facts from an existing knowledge base. The classifier can also provide consistency checking on a knowledge base (which in the case of KL-ONE languages is also referred to as an Ontology).

Another area of knowledge representation research was the problem of common sense reasoning. One of the first realizations from trying to make software that can function with human natural language was that humans regularly draw on an extensive foundation of knowledge about the real world that we simply take for granted but that is not at all obvious to an artificial agent. Basic principles of common sense physics, causality, intentions, etc. An example is the frame problem, that in an event driven logic there need to be axioms that state things maintain position from one moment to the next unless they are moved by some external force. In order to make a true artificial intelligence agent that can converse with humans using natural language and can process basic statements and questions about the world it is essential to represent this kind of knowledge. One of the most ambitious programs to tackle this problem was Doug Lenat's Cyc project. Cyc established its own Frame language and had large numbers of analysts document various areas of common sense reasoning in that language. The knowledge recorded in Cyc included common sense models of time, causality, physics, intentions, and many others.

The starting point for knowledge representation is the *knowledge representation hypothesis* first formalized by Brian C. Smith in 1985:

Any mechanically embodied intelligent process will be comprised of structural ingredients that a) we as external observers naturally take to represent a propositional account of the knowledge that the overall process exhibits, and b) independent of such external semantic attribution, play a formal but causal and essential role in engendering the behavior that manifests that knowledge.

Currently one of the most active areas of knowledge representation research are projects associated with the semantic web. The semantic web seeks to add a layer of semantics (meaning) on top of the current Internet. Rather than indexing web sites and pages via keywords, the semantic web creates large ontologies of concepts. Searching for a concept will be more effective than traditional text only searches. Frame languages and automatic classification play a big part in the vision for the future semantic web. The automatic classification gives developers technology to provide order on a constantly evolving network of knowledge. Defining ontologies that are static and incapable of evolving on the fly would be very limiting for Internet-based systems. The classifier technology provides the ability to deal with the dynamic environment of the Internet.

Recent projects funded primarily by the Defense Advanced Research Projects Agency (DARPA) have integrated frame languages and classifiers with markup languages based on XML. The Resource Description Framework (RDF) provides the basic capability to define classes, subclasses, and properties of objects. The Web Ontology Language (OWL) provides additional levels of semantics and enables integration with classification engines.

Overview

Knowledge-representation is the field of artificial intelligence that focuses on designing computer representations that capture information about the world that can be used to solve complex problems. The justification for knowledge representation is that conventional procedural code is not the best formalism to use to solve complex problems. Knowledge representation makes complex software easier to define and maintain than procedural code and can be used in expert systems.

For example, talking to experts in terms of business rules rather than code lessens the semantic gap between users and developers and makes development of complex systems more practical.

Knowledge representation goes hand in hand with automated reasoning because one of the main purposes of explicitly representing knowledge is to be able to reason about that knowledge, to make inferences, assert new knowledge, etc. Virtually all knowledge representation languages have a reasoning or inference engine as part of the system.

A key trade-off in the design of a knowledge representation formalism is that between expressivity and practicality. The ultimate knowledge representation formalism in terms of expressive power and compactness is First Order Logic (FOL). There is no more powerful formalism than that used by mathematicians to define general propositions about the world. However, FOL has two drawbacks as a knowledge representation formalism: ease of use and practicality of implementation. First order logic can be intimidating even for many software developers. Languages which do not have the complete formal power of FOL can still provide close to the same expressive power with a user interface that is more practical for the average developer to understand. The issue of practicality of implementation is that FOL in some ways is too expressive. With FOL it is possible to create statements (e.g. quantification over infinite sets) that would cause a system to never terminate if it attempted to verify them.

Thus, a subset of FOL can be both easier to use and more practical to implement. This was a driving motivation behind rule-based expert systems. IF-THEN rules provide a subset of FOL but a very useful one that is also very intuitive. The history of most of the early AI knowledge representation formalisms; from databases to semantic nets to theorem provers and production systems can be viewed as various design decisions on whether to emphasize expressive power or computability and efficiency.

In a key 1993 paper on the topic, Randall Davis of MIT outlined five distinct roles to analyze a knowledge representation framework:

- A knowledge representation (KR) is most fundamentally a surrogate, a substitute for the thing itself, used to enable an entity to determine consequences by thinking rather than acting, i.e., by reasoning about the world rather than taking action in it.

- It is a set of ontological commitments, i.e., an answer to the question: In what terms should I think about the world?

- It is a fragmentary theory of intelligent reasoning, expressed in terms of three components: (i) the representation's fundamental conception of intelligent reasoning; (ii) the set of inferences the representation sanctions; and (iii) the set of inferences it recommends.

- It is a medium for pragmatically efficient computation, i.e., the computational environment in which thinking is accomplished. One contribution to this pragmatic efficiency is supplied by the guidance a representation provides for organizing information so as to facilitate making the recommended inferences.

- It is a medium of human expression, i.e., a language in which we say things about the world."

Knowledge representation and reasoning are a key enabling technology for the Semantic web. Languages based on the Frame model with automatic classification provide a layer of semantics on top of the existing Internet. Rather than searching via text strings as is typical today it will be possible to define logical queries and find pages that map to those queries. The automated reasoning component in these systems is an engine known as the classifier. Classifiers focus on the subsumption relations in a knowledge base rather than rules. A classifier can infer new classes and dynamically change the ontology as new information becomes available. This capability is ideal for the ever changing and evolving information space of the Internet.

The Semantic web integrates concepts from knowledge representation and reasoning with markup languages based on XML. The Resource Description Framework (RDF) provides the basic capabilities to define knowledge-based objects on the Internet with basic features such as Is-A relations and object properties. The Web Ontology Language (OWL) adds additional semantics and integrates with automatic classification reasoners.

Characteristics

In 1985, Ron Brachman categorized the core issues for knowledge representation as follows:

- Primitives. What is the underlying framework used to represent knowledge? Semantic networks were one of the first knowledge representation primitives. Also, data structures and algorithms for general fast search. In this area there is a strong overlap with research in data structures and algorithms in computer science. In early systems the Lisp programming language which was modeled after the lambda calculus was often used as a form of functional knowledge representation. Frames and Rules were the next kind of primitive. Frame languages had various mechanisms for expressing and enforcing constraints on frame data. All data in frames are stored in slots. Slots are analogous to relations in entity-relation modeling and to object properties in object-oriented modeling. Another technique for primitives is to define languages that are modeled after First Order Logic (FOL). The most well known example is Prolog but there are also many special purpose theorem proving environments. These environments can validate logical models and can deduce new theories from existing models. Essentially they automate the process a logician would go through in analyzing a model. Theorem proving technology had some specific practical applications in the areas of software engineering. For example, it is possible to prove that a software program rigidly adheres to a formal logical specification.

- Meta-representation. This is also known as the issue of reflection in computer science. It refers to the capability of a formalism to have access to information about its own state. An example would be the meta-object protocol in Smalltalk and CLOS that gives developers run time access to the class objects and enables them to dynamically redefine the structure of the knowledge base even at run time. Meta-representation means the knowledge representation language is itself expressed in that language. For example, in most Frame based environments all frames would be instances of a frame class. That class object can be inspected at run time so that the object can understand and even change its internal structure or the structure of other parts of the model. In rule-based environments the rules were also usually instances of rule classes. Part of the meta protocol for rules were the meta rules that prioritized rule firing.

- Incompleteness. Traditional logic requires additional axioms and constraints to deal with the real world as opposed to the world of mathematics. Also, it is often useful to associate degrees of confidence with a statement. I.e., not simply say "Socrates is Human" but rather "Socrates is Human with confidence 50%". This was one of the early innovations from expert systems research which migrated to some commercial tools, the ability to associate certainty factors with rules and conclusions. Later research in this area is known as Fuzzy Logic.

- Definitions and universals vs. facts and defaults. Universals are general statements about the world such as "All humans are mortal". Facts are specific examples of universals such as "Socrates is a human and therefore mortal". In logical terms definitions and universals are about universal quantification while facts and defaults are about existential quantifications. All forms of knowledge representation must deal with this aspect and most do so with some variant of set theory, modeling universals as sets and subsets and definitions as elements in those sets.

- Non-monotonic reasoning. Non-monotonic reasoning allows various kinds of hypothetical reasoning. The system associates facts asserted with the rules and facts used to justify them and as those facts change updates the dependent knowledge as well. In rule based systems this capability is known as a truth maintenance system.

- Expressive adequacy. The standard that Brachman and most AI researchers use to measure expressive adequacy is usually First Order Logic (FOL). Theoretical limitations mean that a full implementation of FOL is not practical. Researchers should be clear about how expressive (how much of full FOL expressive power) they intend their representation to be.

- Reasoning efficiency. This refers to the run time efficiency of the system. The ability of the knowledge base to be updated and the reasoner to develop new inferences in a reasonable period of time. In some ways this is the flip side of expressive adequacy. In general the more powerful a representation, the more it has expressive adequacy, the less efficient its automated reasoning engine will be. Efficiency was often an issue, especially for early applications of knowledge representation technology. They were usually implemented in interpreted environments such as Lisp which were slow compared to more traditional platforms of the time.

Ontology Engineering

In the early years of knowledge-based systems the knowledge-bases were fairly small. The knowledge-bases that were meant to actually solve real problems rather than do proof of concept demonstrations needed to focus on well defined problems. So for example, not just medical diagnosis as a whole topic but medical diagnosis of certain kinds of diseases.

As knowledge-based technology scaled up the need for larger knowledge bases and for modular knowledge bases that could communicate and integrate with each other became apparent. This gave rise to the discipline of ontology engineering, designing and building large knowledge bases that could be used by multiple projects. One of the leading research projects in this area was the Cyc project. Cyc was an attempt to build a huge encyclopedic knowledge base that would contain not just expert knowledge but common sense knowledge. In designing an artificial intelligence agent it was soon realized that representing common sense knowledge, knowledge that humans simply take for granted, was essential to make an AI that could interact with humans using natural language. Cyc was meant to address this problem. The language they defined was known as CycL.

After CycL, a number of ontology languages have been developed. Most are declarative languages, and are either frame languages, or are based on first-order logic. Modularity—the ability to define boundaries around specific domains and problem spaces—is essential for these languages because as stated by Tom Gruber, "Every ontology is a treaty- a social agreement among people with common motive in sharing." There are always many competing and differing views that make any general purpose ontology impossible. A general purpose ontology would have to be applicable in any domain and different areas of knowledge need to be unified.

There is a long history of work attempting to build ontologies for a variety of task domains, e.g., an ontology for liquids, the lumped element model widely used in representing electronic circuits (e.g.,), as well as ontologies for time, belief, and even programming itself. Each of these offers a way to see some part of the world. The lumped element model, for instance, suggests that we think of circuits in terms of components with connections between them, with signals flowing instantaneously along the connections. This is a useful view, but not the only possible one. A different ontology arises if we need to attend to the electrodynamics in the device: Here signals propagate at finite speed and an object (like a resistor) that was previously viewed as a single component with an I/O behavior may now have to be thought of as an extended medium through which an electromagnetic wave flows.

Ontologies can of course be written down in a wide variety of languages and notations (e.g., logic, LISP, etc.); the essential information is not the form of that language but the content, i.e., the set of concepts offered as a way of thinking about the world. Simply put, the important part is notions like connections and components, not the choice between writing them as predicates or LISP constructs.

The commitment made selecting one or another ontology can produce a sharply different view of the task at hand. Consider the difference that arises in selecting the lumped element view of a circuit rather than the electrodynamic view of the same device. As a second example, medical diagnosis viewed in terms of rules (e.g., MYCIN) looks substantially different from the same task viewed in terms of frames (e.g., INTERNIST). Where MYCIN sees the medical world as made up of empirical associations connecting symptom to disease, INTERNIST sees a set of prototypes, in particular prototypical diseases, to be matched against the case at hand.

Intelligent agents should have capacity for:

- Perceiving, that is, acquiring information from environment,

- Knowledge Representation, that is, representing its understanding of the world,

- Reasoning, that is, inferring the implications of what it knows and of the choices it has, and

- Acting, that is, choosing what it want to do and carry it out.

Representation of knowledge and the reasoning process are central to the entire field of artificial intelligence. The primary component of a knowledge-based agent is its knowledge-base. A knowledge-base is a set of sentences. Each sentence is expressed in a language called the knowledge representation language. Sentences represent some assertions about the world. There must mechanisms to derive new sentences from old ones. This process is known as inferencing or reasoning.

Inference must obey the primary requirement that the new sentences should follow logically from the previous ones. *Logic* is the primary vehicle for representing and reasoning about knowledge. Specifically, we will be dealing with formal logic. The advantage of using formal logic as a language of AI is that it is precise and definite. This allows programs to be written which are declarative - they describe what is true and not how to solve problems. This also allows for automated reasoning techniques for general purpose inferencing. This, however, leads to some severe limitations. Clearly, a large portion of the reasoning carried out by humans depends on handling knowledge that is uncertain. Logic cannot represent this uncertainty well. Similarly, natural language reasoning requires inferring hidden state, namely, the intention of the speaker. When we say, "One of the wheel of the car is flat.", we know that it has three wheels left. Humans can cope with virtually infinite variety of utterances using a finite store of commonsense knowledge. Formal logic has difficulty with this kind of ambiguity.

A logic consists of two parts, a language and a method of reasoning. The logical language, in turn, has two aspects, syntax and semantics. Thus, to specify or define a particular logic, one needs to specify three things:

Syntax: The atomic symbols of the logical language, and the rules for constructing well-formed, non-atomic expressions (symbol structures) of the logic. Syntax specifies the symbols in the language and how they can be combined to form sentences. Hence facts about the world are represented as sentences in logic.

Semantics: The meanings of the atomic symbols of the logic, and the rules for determining the meanings of non-atomic expressions of the logic. It specifies what facts in the world a sentence refers to. Hence, also specifies how you assign a truth value to a sentence based on its meaning in the world. A fact is a claim about the world, and may be true or false.

Syntactic Inference Method: The rules for determining a subset of logical expressions, called theorems of the logic. It refers to mechanical method for computing (deriving) new (true) sentences from existing sentences.

Facts are claims about the world that are True or False, whereas a representation is an expression (sentence) in some language that can be encoded in a computer program and stands for the objects and relations in the world. We need to ensure that the representation is consistent with reality, so that the following figure holds:

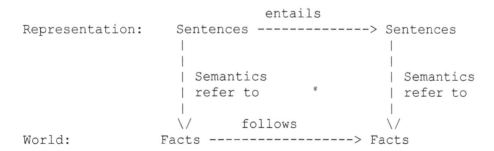

There are a number of logical systems with different syntax and semantics. Listed below a few of them.

- Propositional logic

    All objects described are fixed or unique

    "John is a student" student(john)

    Here John refers to one unique person.

- First order predicate logic

    Objects described can be unique or variables to stand for a unique object

    "All students are poor"

    ForAll(S) [student(S) -> poor(S)]

    Here S can be replaced by many different unique students.

    This makes programs much more compact:

        eg. ForAll(A,B)[brother(A,B) -> brother (B,A)]

        replaces half the possible statements about brothers
- Temporal

    Represents truth over time.

- Modal

    Represents doubt

- Higher order logics

    Allows variable to represent many relations between objects

- Non-monotonic

    Represents defaults

Propositional is one of the simplest systems of logic.

Propositional Logic

In propositional logic (PL) an user defines a set of propositional symbols, like P and Q. User defines the semantics of each of these symbols. For example,

    o P means "It is hot"

    o Q means "It is humid"

    o R means "It is raining"

- A sentence (also called a formula or well-formed formula or wff) is defined as:

  1. A symbol

  2. If S is a sentence, then ~S is a sentence, where "~" is the "not" logical operator

  3. If S and T are sentences, then (S v T), (S ^ T), (S => T), and (S <=> T) are sentences, where the four logical connectives correspond to "or," "and," "implies," and "if and only if," respectively

  4. A finite number of applications of (1)-(3)

- Examples of PL sentences:

  o (P ^ Q) => R (here meaning "If it is hot and humid, then it is raining")

  o Q => P (here meaning "If it is humid, then it is hot")

  o Q (here meaning "It is humid.")

- Given the truth values of all of the constituent symbols in a sentence, that sentence can be "evaluated" to determine its truth value (True or False). This is called an interpretation of the sentence.

- A model is an interpretation (i.e., an assignment of truth values to symbols) of a set of sentences such that each sentence is True. A model is just a formal mathematical structure that "stands in" for the world.

- A valid sentence (also called a tautology) is a sentence that is True under *all* interpretations. Hence, no matter what the world is actually like or what the semantics is, the sentence is True. For example "It's raining or it's not raining."

- An inconsistent sentence (also called unsatisfiable or a contradiction) is a sentence that is False under *all* interpretations. Hence the world is never like what it describes. For example, "It's raining and it's not raining."

- Sentence P entails sentence Q, written P |= Q, means that whenever P is True, so is Q. In other words, all models of P are also models of Q

  Entailment ($\models$) : Given 2 sentences p and q we say p entails q, written $p \models q$, if q holds in every model that p holds

Example: Entailment

$$p \wedge (p \Rightarrow q) \models q$$

Show that:

Proof: For any model M in which $p \wedge (p \Rightarrow q)$ holds then we know that p holds in M and $p \Rightarrow q$ holds in M. Since p holds in M then since $p \Rightarrow q$ holds in M, q must hold in M. Therefore q holds in every model that $p \wedge (p \Rightarrow q)$ holds and so $p \wedge (p \Rightarrow q) \models q$.

As we have noted models affect equivalence and so we repeat the definition again and give an example of a proof of equivalence.

  Equivalnce: Two sentences are equivalent if they hold in exactly the same models.

Example: Equivalence

$$p \Rightarrow q \equiv -p \vee q$$

Show that:

Proof: We need to provide two proofs as above for

$$p \Rightarrow q \models - p \vee q$$

- For any model M in which $p \Rightarrow q$ holds then we know that either p holds in M and so q holds in M, or p does not hold in M and so $-p$ holds in M. Since either q holds in M or $-p$ holds in M, then $-p \vee q$ holds in M.

and $-p \vee q \models p \Rightarrow q$

- For any model M in which $-p \vee q$ holds then we know that either $-p$ holds in M or q holds in M. If $-p$ holds in M then $p \Rightarrow q$ holds in M. Otherwise, if q holds in M then $p \Rightarrow q$ holds in M. Therefore $p \Rightarrow q$ holds in M.

Knowledge based programming relies on concluding new knowledge from existing knowledge. Entailment is a required justification; i.e. if $P_1, \cdots \cdots P_n$ is known then there is justification to conclude q if

$$P_1 \wedge .... \wedge P_n \models q$$

In some circumstances we insist on this strong form of justification; i.e. we cannot conclude q unless the entailment holds. Reasoning like this is the equivalent for knowledge based programs of running a piece of conventional software.

Note: Entailment ($\models$) is concerned with *truth* and is determined by considering the truth of the sentences in all models.

Propositional Logic Inference

Let KB = { S1, S2,..., SM } be the set of all sentences in our Knowledge Base, where each Si is a sentence in Propositional Logic. Let { X1, X2, ..., XN } be the set of all the symbols (i.e., variables) that are contained in all of the M sentences in KB. Say we want to know if a goal (aka query, conclusion, or theorem) sentence G follows from KB.

Model Checking

```
X1 X2 ... XN | S1 S2 ... SM | S1 ^ S2 ^...^ SM | G | (S1 ^...^ SM) => G
-------------|--------------|------------------|---|-------------------
F  F  ... F  |              |                  |   |
F  F  ... T  |              |                  |   |
...          |              |                  |   |
T  T  ... T  |              |                  |   |
```

Since the computer doesn't know the interpretation of these sentences in the world, we don't know whether the constituent symbols represent facts in the world that are True or False. So, instead, consider *all* possible combinations of truth values for all the symbols, hence enumerating all logically distinct cases:

- There are 2^N rows in the table.

- Each row corresponds to an equivalence class of worlds that, under a given interpretation, have the truth values for the N symbols assigned in that row.

- The models of KB are the rows where the third-to-last column is *true*, i.e., where all of the sentences in KB are *true*.

- A sentence R is valid if and only if it is true under all possible interpretations, i.e., if the entire column associated with R contains all *true* values.

- Since we don't know the semantics and therefore whether each symbol is True or False, to determine if a sentence G is entailed by KB, we must determine if all models of KB are also models of G. That is, whenever KB is true, G is true too. In other words, whenever the third-to-last column has a T, the same row in the second-to-last column also has a T. But this is logically equivalent to saying that the sentence (KB => G) is valid (by definition of the "implies" connective). In other words, if the last column of the table above contains only *True*, then KB entails G; or conclusion G logically follows from the premises in KB, no matter what the interpretations (i.e., semantics) associated with all of the sentences!

- The truth table method of inference is complete for PL (Propositional Logic) because we can always enumerate all 2^n rows for the n propositional symbols that occur. But this is exponential in n. In general, it has been shown that the problem of checking if a set of sentences in PL is satisfiable is NP-complete. (The truth table method of inference is *not* complete for FOL (First-Order Logic).)

Example

Using the "weather" sentences from above, let KB = $(((P \wedge Q) => R) \wedge (Q => P) \wedge Q)$

corresponding to the three facts we know about the weather: (1) "If it is hot and humid, then it is raining," (2) "If it is humid, then it is hot," and (3) "It is humid." Now let's ask the query "Is it raining?" That is, is the query sentence R entailed by KB? Using the truth-table approach to answering this query we have:

P Q R	(P ^ Q) => R	Q => P	Q	KB	R	KB => R
T T T	T	T	T	T	T	T
T T F	F	T	T	F	F	T
T F T	T	T	F	F	T	T
T F F	T	T	F	F	F	T
F T T	T	F	T	F	T	T
F T F	T	F	T	F	F	T
F F T	T	T	F	F	T	T
F F F	T	T	F	F	F	T

Hence, in this problem there is only one model of KB, when P, Q, and R are all True. And in this case R is also True, so R is entailed by KB. Also, you can see that the last column is all True values, so the sentence KB => R is valid.

Instead of an exponential length proof by truth table construction, is there a faster way to implement the inference process? Yes, using a proof procedure or inference procedure that uses sound rules of inference to deduce (i.e., derive) new sentences that are true in all cases where the premises are true. For example, consider the following:

```
P    Q | P    P => Q | P ^ (P => Q) | Q | (P ^ (P => Q)) => Q
------ |------------ |--------------|---|--------------------------
F    F | F     T     |      F       | F |          T
F    T | F     T     |      F       | T |          T
T    F | T     F     |      F       | F |          T
T    T | T     T     |      T       | T |          T
```

Since whenever P and P => Q are both true (last row only), Q is true too, Q is said to be derived from these two premise sentences. We write this as KB |- Q. This local pattern referencing only two of the M sentences in KB is called the Modus Ponens inference rule. The truth table shows that this inference rule is sound. It specifies how to make one kind of step in deriving a conclusion sentence from a KB.

Therefore, given the sentences in KB, construct a proof that a given conclusion sentence can be derived from KB by applying a sequence of sound inferences using either sentences in KB or sentences derived earlier in the proof, until the conclusion sentence is derived. This method is called the Natural Deduction procedure. (Note: This step-by-step, local proof process also relies on the monotonicity property of PL and FOL. That is, adding a new sentence to KB does not affect what can be entailed from the original KB and does not invalidate old sentences.)

Rule of Inference

In logic, a rule of inference, inference rule or transformation rule is a logical form consisting of a function which takes premises, analyzes their syntax, and returns a conclusion (or conclusions). For example, the rule of inference called *modus ponens* takes two premises, one in the form "If p then q" and another in the form "p", and returns the conclusion "q". The rule is valid with respect to the semantics of classical logic (as well as the semantics of many other non-classical logics), in the sense that if the premises are true (under an interpretation), then so is the conclusion.

Typically, a rule of inference preserves truth, a semantic property. In many-valued logic, it preserves a general designation. But a rule of inference's action is purely syntactic, and does not need to preserve any semantic property: any function from sets of formulae to formulae counts as a rule of inference. Usually only rules that are recursive are important; i.e. rules such that there is an effective procedure for determining whether any given formula is the conclusion of a given set of formulae according to the rule. An example of a rule that is not effective in this sense is the infinitary ω-rule.

Popular rules of inference in propositional logic include *modus ponens*, *modus tollens*, and contraposition. First-order predicate logic uses rules of inference to deal with logical quantifiers.

The Standard form of Rules of Inference

In formal logic (and many related areas), rules of inference are usually given in the following standard form:
Premise#1
Premise#2
...
Premise#n
_____
Conclusion

This expression states that whenever in the course of some logical derivation the given premises have been obtained, the specified conclusion can be taken for granted as well. The exact formal language that is used to describe both premises and conclusions depends on the actual context of the derivations. In a simple case, one may use logical formulae, such as in:

$$A \to B$$
$$\frac{A}{B}$$

This is the *modus ponens* rule of propositional logic. Rules of inference are often formulated as schemata employing metavariables. In the rule (schema) above, the metavariables A and B can be instantiated to any element of the universe (or sometimes, by convention, a restricted subset such as propositions) to form an infinite set of inference rules.

A proof system is formed from a set of rules chained together to form proofs, also called *derivations*. Any derivation has only one final conclusion, which is the statement proved or derived. If premises are left unsatisfied in the derivation, then the derivation is a proof of a *hypothetical* statement: "*if* the premises hold, *then* the conclusion holds."

Axiom Schemas and Axioms

Inference rules may also be stated in this form: (1) zero or more premises, (2) a turnstile symbol $\vdash p$, which means "infers", "proves", or "concludes", and (3) a conclusion. This form usually embodies the relational (as opposed to functional) view of a rule of inference, where the turnstile stands for a deducibility relation holding between premises and conclusion.

An inference rule containing no premises is called an axiom schema or, if it contains no metavariables, simply an axiom.

Rules of inference must be distinguished from axioms of a theory. In terms of semantics, axioms are valid assertions. Axioms are usually regarded as starting points for applying rules of inference and generating a set of conclusions. Or, in less technical terms:

Rules are statements *about* the system, axioms are statements *in* the system. For example:

- The rule that from $\vdash p$ we can infer $\vdash$ Provable(p) is a statement that says if you've proven p, it follows that p is provable. This rule holds in Peano arithmetic, for example.

- The axiom $p \rightarrow \text{Provable}(p)$ would mean that every true statement is provable. This axiom does not hold in Peano arithmetic.

Rules of inference play a vital role in the specification of logical calculi as they are considered in proof theory, such as the sequent calculus and natural deduction.

Example: Hilbert systems for two propositional logics

In a Hilbert system, the premises and conclusion of the inference rules are simply formulae of some language, usually employing metavariables. For graphical compactness of the presentation and to emphasize the distinction between axioms and rules of inference, this section uses the sequent notation ($\vdash$) instead of a vertical presentation of rules.

The formal language for classical propositional logic can be expressed using just negation ($\neg$), implication ($\rightarrow$) and propositional symbols. A well-known axiomatization, comprising three axiom schemata and one inference rule (*modus ponens*), is:

(CA1) $\vdash A \rightarrow (B \rightarrow A)$
(CA2) $\vdash (A \rightarrow (B \rightarrow C)) \rightarrow ((A \rightarrow B) \rightarrow (A \rightarrow C))$
(CA3) $\vdash (\neg A \rightarrow \neg B) \rightarrow (B \rightarrow A)$
(MP) $A, A \rightarrow B \vdash B$

It may seem redundant to have two notions of inference in this case, $\vdash$ and $\rightarrow$. In classical propositional logic, they indeed coincide; the deduction theorem states that $A \vdash B$ if and only if $\vdash A \rightarrow B$. There is however a distinction worth emphasizing even in this case: the first notation describes a deduction, that is an activity of passing from sentences to sentences, whereas $A \rightarrow B$ is simply a formula made with a logical connective, implication in this case. Without an inference rule (like *modus ponens* in this case), there is no deduction or inference. This point is illustrated in Lewis Carroll's dialogue called "What the Tortoise Said to Achilles".

For some non-classical logics, the deduction theorem does not hold. For example, the three-valued logic Ł3 of Łukasiewicz can be axiomatized as:

(CA1) $\vdash A \rightarrow (B \rightarrow A)$
(LA2) $\vdash (A \rightarrow B) \rightarrow ((B \rightarrow C) \rightarrow (A \rightarrow C))$
(CA3) $\vdash (\neg A \rightarrow \neg B) \rightarrow (B \rightarrow A)$
(LA4) $\vdash ((A \rightarrow \neg A) \rightarrow A) \rightarrow A$
(MP) $A, A \rightarrow B \vdash B$

This sequence differs from classical logic by the change in axiom 2 and the addition of axiom 4. The classical deduction theorem does not hold for this logic, however a modified form does hold, namely $A \vdash B$ if and only if $\vdash A \rightarrow (A \rightarrow B)$.

Admissibility and Derivability

In a set of rules, an inference rule could be redundant in the sense that it is *admissible* or *derivable*. A derivable rule is one whose conclusion can be derived from its premises using the other rules. An admissible rule is one whose conclusion holds whenever the premises hold. All derivable rules are admissible. To appreciate the difference, consider the following set of rules for defining the natural numbers (the judgment n nat asserts the fact that n is a natural number):

$$\frac{}{0\,\mathsf{nat}} \qquad \frac{n\,\mathsf{nat}}{\mathsf{s}(n)\,\mathsf{nat}}$$

The first rule states that 0 is a natural number, and the second states that s(n) is a natural number if n is. In this proof system, the following rule, demonstrating that the second successor of a natural number is also a natural number, is derivable:

$$\frac{n\,\mathsf{nat}}{\mathsf{s}(\mathsf{s}(n))\,\mathsf{nat}}$$

Its derivation is the composition of two uses of the successor rule above. The following rule for asserting the existence of a predecessor for any nonzero number is merely admissible:

$$\frac{\mathsf{s}(n)\,\mathsf{nat}}{n\,\mathsf{nat}}$$

This is a true fact of natural numbers, as can be proven by induction. (To prove that this rule is admissible, assume a derivation of the premise and induct on it to produce a derivation n nat of .) However, it is not derivable, because it depends on the structure of the derivation of the premise. Because of this, derivability is stable under additions to the proof system, whereas admissibility is not. To see the difference, suppose the following nonsense rule were added to the proof system:

$$\frac{}{\mathsf{s}(-3)\,\mathsf{nat}}$$

In this new system, the double-successor rule is still derivable. However, the rule for finding the predecessor is no longer admissible, because there is no way to derive −3 nat . The brittleness of admissibility comes from the way it is proved: since the proof can induct on the structure of the derivations of the premises, extensions to the system add new cases to this proof, which may no longer hold.

Admissible rules can be thought of as theorems of a proof system. For instance, in a sequent calculus where cut elimination holds, the *cut* rule is admissible.

Here are some examples of sound rules of inference. Each can be shown to be sound once and for all using a truth table. The left column contains the premise sentence(s), and the right column contains the derived sentence. We write each of these derivations as A |- B , where A is the premise and B is the derived sentence.

Name	Premise(s)	Derived Sentence
Modus Ponens	A, A => B	B
And Introduction	A, B	AB
And Elimination	AB	A
Double Negation	~~A	A
Unit Resolution	A v B, ~B	A
Resolution	A v B, ~B v C	A v C

In addition to the above rules of inference one also requires a set of equivalences of propositional logic like "A /\ B" is equivalent to "B /\ A". A number of such equivalences were presented in the discussion on propositional logic.

Using Inference Rules to Prove a Query/Goal/Theorem

A proof is a sequence of sentences, where each sentence is either a premise or a sentence derived from earlier sentences in the proof by one of the rules of inference. The last sentence is the query (also called goal or theorem) that we want to prove.

Example for the "weather problem" given above.

1.	Q	Premise
2.	Q => P	Premise
3.	P	Modus Ponens(1,2)
4.	(PQ) => R	Premise
5.	PQ	And Introduction(1,3)
6.	R	Modus Ponens(4,5)

Inference vs Entailmant

There is a subtle difference between entailment and inference.

Inference ($\models$) : Given 2 sentences p and q we say q is inferred from p, written $p \vdash q$, if there is a sequence of rules of inference that apply to p and allow q to be added.

Notice that inference is not directly related to truth; i.e. we can infer a sentence provided we have rules of inference that produce the sentence from the original sentences.

However, if rules of inference are to be useful we wish them to be related to entailment. Ideally we would like:

$$p \vdash q_{iff} \; p \models q$$

but this equivalence may fail in two ways:

$$p \vdash q_{but} \; p \not\models q$$

We have inferred q by applying rules of inference to p, but there is some model in which p holds but q does not hold. In this case the rules of inference have inferred ``too much".

$$p \models q_{but} \; p \not\vdash q$$

q is a sentence which holds in all models in which p holds, but we cannot find rules of inference that will infer q from p. In this case the rules of inference are insufficient to infer the things we want to be able to infer.

Soundness and Completeness

These notions are so important that there are 2 properties of logics associated with them.

> Soundness : An inference procedure $|-$ is sound if whenever $p|-q$ then it is also the case that $p|=q$.

``A sound inference procedure infers things that are valid consequences"

> Completeness : An inference procedure $|-$ is complete if whenever $p|=q$ then it is also the case that $p|-q$.

> "A complete inference procedure is able to infer anything that is that is a valid consequence"

The ``best" inference procedures are both sound and complete, but gaining completeness is often computationally expensive. Notice that even if inference is not complete it is desirable that it is sound.

Propositional Logic and Predicate Logic each with Modus Ponens as their inference produce are sound but not complete. We shall see that we need further (sound) rules of inference to achieve completeness. In fact we shall see that we shall even restrict the language in order to achieve an effective inference procedure that is sound and complete for a subset of First Order Predicate Logic.

The notion of soundness and completeness is more generally applied than in logic. Whenever we create a knowledge based program we use the syntax of the knowledge representation language, we assign a semantics in some way and the reasoning mechanism defines the inference procedures. The semantics will define what entailment means in this representation and we will be interested in how well the reasoning mechanism achieves entailment.

Decidability

Determining whether $p|=q$ is computationally hard. If q is a consequence then if $|-$ is complete then we know that $p|-q$ and we can apply the rules of inference exhaustively knowing that we will eventually find the sequence of rules of inference. It may take a long time but it is finite.

However if q is not a consequence (remember the task is *whether or not* $p|=q$) then we can happily apply rules of inference generating more and more irrelevant consequences. So the procedure is guaranteed to eventually stop if q is derivable, but may not terminate otherwise.

> Decidability : A problem is decidable if there is a procedure that is guaranteed to terminate having determined whether the answer is "yes" or "no".

> > Semi $-$ Decidability : A problem is only semi $-$ decidable if there is a procedure that is guaranteed to terminate in one of these cases but not both.

Entailment in Propositional Logic is decidable since truth tables can be applied in a finite number of steps to determine whether or not $p|=q$.

Entailment in Predicate Logic is only semi-decidable; it is only guaranteed to terminate when q is a consequence. One result of this semi-decidability is that many problems are not decidable; if they rely on failing to prove some sentence. Planning is typically not decidable. A common reaction to a non-decidable problem is to *assume* the answer after some reasoning time threshold has been reached. Another reaction to the semi-decidability of Predicate Logic is to restrict attention to subsets of the logic; however even if its entailment is decidable the procedure may be computationally expensive.

First-order Logic

First-order logic – also known as first-order predicate calculus and predicate logic – is a collection of formal systems used in mathematics, philosophy, linguistics, and computer science. First-order logic uses quantified variables over non-logical objects and allows the use of sentences that contain variables, so that rather than propositions such as *Socrates is a man* one can have expressions in the form "there exists X such that X is Socrates and X is a man" where *there exists* is a quantifier and X is a variable. This distinguishes it from propositional logic, which does not use quantifiers.

A theory about a topic is usually a first-order logic together with a specified domain of discourse over which the quantified variables range, finitely many functions from that domain to itself, finitely many predicates defined on that domain, and a set of axioms believed to hold for those things. Sometimes "theory" is understood in a more formal sense, which is just a set of sentences in first-order logic.

The adjective "first-order" distinguishes first-order logic from higher-order logic in which there are predicates having predicates or functions as arguments, or in which one or both of predicate quantifiers or function quantifiers are permitted. In first-order theories, predicates are often associated with sets. In interpreted higher-order theories, predicates may be interpreted as sets of sets.

There are many deductive systems for first-order logic which are both sound (all provable statements are true in all models) and complete (all statements which are true in all models are provable). Although the logical consequence relation is only semidecidable, much progress has been made in automated theorem proving in first-order logic. First-order logic also satisfies several metalogical theorems that make it amenable to analysis in proof theory, such as the Löwenheim–Skolem theorem and the compactness theorem.

First-order logic is the standard for the formalization of mathematics into axioms and is studied in the foundations of mathematics. Peano arithmetic and Zermelo–Fraenkel set theory are axiomatizations of number theory and set theory, respectively, into first-order logic. No first-order theory, however, has the strength to uniquely describe a structure with an infinite domain, such as the natural numbers or the real line. Axioms systems that do fully describe these two structures (that is, categorical axiom systems) can be obtained in stronger logics such as second-order logic.

Introduction

While propositional logic deals with simple declarative propositions, first-order logic additionally covers predicates and quantification.

A predicate takes an entity or entities in the domain of discourse as input and outputs either True or False. Consider the two sentences "Socrates is a philosopher" and "Plato is a philosopher". In propositional logic, these sentences are viewed as being unrelated and might be denoted, for example, by variables such as p and q. The predicate "is a philosopher" occurs in both sentences, which have a common structure of "a is a philosopher". The variable a is instantiated as "Socrates" in the first sentence and is instantiated as "Plato" in the second sentence. While first-order logic allows for the use of predicates, such as "is a philosopher" in this example, propositional logic does not.

Relationships between predicates can be stated using logical connectives. Consider, for example, the first-order formula "if a is a philosopher, then a is a scholar". This formula is a conditional statement with "a is a philosopher" as its hypothesis and "a is a scholar" as its conclusion. The truth of this formula depends on which object is denoted by a, and on the interpretations of the predicates "is a philosopher" and "is a scholar".

Quantifiers can be applied to variables in a formula. The variable a in the previous formula can be universally quantified, for instance, with the first-order sentence "For every a, if a is a philosopher, then a is a scholar". The universal quantifier "for every" in this sentence expresses the idea that the claim "if a is a philosopher, then a is a scholar" holds for *all* choices of a.

The *negation* of the sentence "For every a, if a is a philosopher, then a is a scholar" is logically equivalent to the sentence "There exists a such that a is a philosopher and a is not a scholar". The existential quantifier "there exists" expresses the idea that the claim "a is a philosopher and a is not a scholar" holds for *some* choice of a.

The predicates "is a philosopher" and "is a scholar" each take a single variable. In general, predicates can take several variables. In the first-order sentence "Socrates is the teacher of Plato", the predicate "is the teacher of" takes two variables.

An interpretation (or model) of a first-order formula specifies what each predicate means and the entities that can instantiate the variables. These entities form the domain of discourse or universe, which is usually required to be a nonempty set. For example, in an interpretation with the domain of discourse consisting of all human beings and the predicate "is a philosopher" understood as "was the author of the *Republic*", the sentence "There exists a such that a is a philosopher" is seen as being true, as witnessed by Plato.

Syntax

There are two key parts of first-order logic. The syntax determines which collections of symbols are legal expressions in first-order logic, while the semantics determine the meanings behind these expressions.

Alphabet

Unlike natural languages, such as English, the language of first-order logic is completely formal, so that it can be mechanically determined whether a given expression is legal. There are two key types of legal expressions: terms, which intuitively represent objects, and formulas, which intuitively express predicates that can be true or false. The terms and formulas of first-order logic are strings of symbols which together form the alphabet of the language. As with all formal languages, the nature of the symbols themselves is outside the scope of formal logic; they are often regarded simply as letters and punctuation symbols.

It is common to divide the symbols of the alphabet into logical symbols, which always have the same meaning, and non-logical symbols, whose meaning varies by interpretation. For example, the logical symbol $\wedge$ always represents "and"; it is never interpreted as "or". On the other hand, a non-logical predicate symbol such as Phil(x) could be interpreted to mean "x is a philosopher", "x is a man named Philip", or any other unary predicate, depending on the interpretation at hand.

Logical symbols

There are several logical symbols in the alphabet, which vary by author but usually include:

- The quantifier symbols $\forall$ and $\exists$

- The logical connectives: $\wedge$ for conjunction, $\vee$ for disjunction, $\rightarrow$ for implication, $\leftrightarrow$ for biconditional, $\neg$ for negation. Occasionally other logical connective symbols are included. Some authors use Cpq, instead of $\rightarrow$, and Epq, instead of $\leftrightarrow$, especially in contexts where $\rightarrow$ is used for other purposes. Moreover, the horseshoe $\supset$ may replace $\rightarrow$; the triple-bar $\equiv$ may replace $\leftrightarrow$; a tilde ($\sim$), Np, or Fpq, may replace $\neg$; $||$, or Apq may replace $\vee$; and &, Kpq, or the middle dot, $\cdot$, may replace $\wedge$, especially if these symbols are not available for technical reasons. (*Note*: the aforementioned symbols Cpq, Epq, Np, Apq, and Kpq are used in Polish notation.)

- Parentheses, brackets, and other punctuation symbols. The choice of such symbols varies depending on context.

- An infinite set of variables, often denoted by lowercase letters at the end of the alphabet x, y, z, ... Subscripts are often used to distinguish variables: x_0, x_1, x_2, ...

- An equality symbol (sometimes, identity symbol).

It should be noted that not all of these symbols are required – only one of the quantifiers, negation and conjunction, variables, brackets and equality suffice. There are numerous minor variations that may define additional logical symbols:

- Sometimes the truth constants T, Vpq, or $\top$, for "true" and F, Opq, or $\bot$, for "false" are included. Without any such logical operators of valence 0, these two constants can only be expressed using quantifiers.

- Sometimes additional logical connectives are included, such as the Sheffer stroke, Dpq (NAND), and exclusive or, Jpq.

Non-logical Symbols

The non-logical symbols represent predicates (relations), functions and constants on the domain of discourse. It used to be standard practice to use a fixed, infinite set of non-logical symbols for all purposes. A more recent practice is to use different non-logical symbols according to the application one has in mind. Therefore, it has become necessary to name the set of all non-logical symbols used in a particular application. This choice is made via a signature.

The traditional approach is to have only one, infinite, set of non-logical symbols (one signature) for all applications. Consequently, under the traditional approach there is only one language of first-order logic. This approach is still common, especially in philosophically oriented books.

1. For every integer $n \geq 0$ there is a collection of n-ary, or n-place, predicate symbols. Because they represent relations between n elements, they are also called relation symbols. For each arity n we have an infinite supply of them:

$$P^n_0, P^n_1, P^n_2, P^n_3, \ldots$$

2. For every integer $n \geq 0$ there are infinitely many n-ary function symbols:

$$f^n_0, f^n_1, f^n_2, f^n_3, \ldots$$

In contemporary mathematical logic, the signature varies by application. Typical signatures in mathematics are $\{1, \times\}$ or just $\{\times\}$ for groups, or $\{0, 1, +, \times, <\}$ for ordered fields. There are no restrictions on the number of non-logical symbols. The signature can be empty, finite, or infinite, even uncountable. Uncountable signatures occur for example in modern proofs of the Löwenheim-Skolem theorem.

In this approach, every non-logical symbol is of one of the following types.

1. A predicate symbol (or relation symbol) with some valence (or arity, number of arguments) greater than or equal to 0. These are often denoted by uppercase letters P, Q, R,... .

   o Relations of valence 0 can be identified with propositional variables. For example, P, which can stand for any statement.

   o For example, $P(x)$ is a predicate variable of valence 1. One possible interpretation is "x is a man".

   o $Q(x,y)$ is a predicate variable of valence 2. Possible interpretations include "x is greater than y" and "x is the father of y".

2. A function symbol, with some valence greater than or equal to 0. These are often denoted by lowercase letters f, g, h,... .

   o Examples: $f(x)$ may be interpreted as for "the father of x". In arithmetic, it may stand for "-x". In set theory, it may stand for "the power set of x". In arithmetic, $g(x,y)$ may stand for "$x+y$". In set theory, it may stand for "the union of x and y".

   o Function symbols of valence 0 are called constant symbols, and are often denoted by lowercase letters at the beginning of the alphabet a, b, c,... . The symbol a may stand for Socrates. In arithmetic, it may stand for 0. In set theory, such a constant may stand for the empty set.

The traditional approach can be recovered in the modern approach by simply specifying the "custom" signature to consist of the traditional sequences of non-logical symbols.

Formation Rules

The formation rules define the terms and formulas of first order logic. When terms and formulas are represented as strings of symbols, these rules can be used to write a formal grammar for terms and formulas. These rules are generally context-free (each production has a single symbol on the left side), except that the set of symbols may be allowed to be infinite and there may be many start symbols, for example the variables in the case of terms.

Terms

The set of terms is inductively defined by the following rules:

1. Variables. Any variable is a term.

2. Functions. Any expression $f(t_1,...,t_n)$ of n arguments (where each argument t_i is a term and f is a function symbol of valence n) is a term. In particular, symbols denoting individual constants are 0-ary function symbols, and are thus terms.

Only expressions which can be obtained by finitely many applications of rules 1 and 2 are terms. For example, no expression involving a predicate symbol is a term.

Formulas

The set of formulas (also called well-formed formulas or wffs) is inductively defined by the following rules:

1. Predicate symbols. If P is an n-ary predicate symbol and t_1, ..., t_n are terms then $P(t_1,...,t_n)$ is a formula.

2. Equality. If the equality symbol is considered part of logic, and t_1 and t_2 are terms, then $t_1 = t_2$ is a formula.

3. Negation. If φ is a formula, then $\neg \varphi$ is a formula.

4. Binary connectives. If φ and ψ are formulas, then $(\varphi \rightarrow \psi)$ is a formula. Similar rules apply to other binary logical connectives.

5. Quantifiers. If φ is a formula and x is a variable, then $\forall x\varphi$ (for all x, φ holds) and $\exists x\varphi$ (there exists x such that φ) are formulas.

Only expressions which can be obtained by finitely many applications of rules 1–5 are formulas. The formulas obtained from the first two rules are said to be atomic formulas.

For example,

$$\forall x \forall y (P(f(x)) \rightarrow \neg(P(x) \rightarrow Q(f(y),x,z)))$$

is a formula, if f is a unary function symbol, P a unary predicate symbol, and Q a ternary predicate symbol. On the other hand, $\forall x x \rightarrow$ is not a formula, although it is a string of symbols from the alphabet.

The role of the parentheses in the definition is to ensure that any formula can only be obtained in one way by following the inductive definition (in other words, there is a unique parse tree for each formula). This property is known as unique readability of formulas. There are many conventions for where parentheses are used in formulas. For example, some authors use colons or full stops instead of parentheses, or change the places in which parentheses are inserted. Each author's particular definition must be accompanied by a proof of unique readability.

This definition of a formula does not support defining an if-then-else function ite(c, a, b), where "c" is a condition expressed as a formula, that would return "a" if c is true, and "b" if it is false. This is because both predicates and functions can only accept terms as parameters, but the first parameter is a formula. Some languages built on first-order logic, such as SMT-LIB 2.0, add this.

Notational Conventions

For convenience, conventions have been developed about the precedence of the logical operators, to avoid the need to write parentheses in some cases. These rules are similar to the order of operations in arithmetic. A common convention is:

- $\neg$ is evaluated first

- $\wedge$ and $\vee$ are evaluated next

- Quantifiers are evaluated next

- $\rightarrow$ is evaluated last.

Moreover, extra punctuation not required by the definition may be inserted to make formulas easier to read. Thus the formula:

$$(\neg \forall x P(x) \rightarrow \exists x \neg P(x))$$

might be written as:

$$(\neg[\forall x P(x)]) \rightarrow \exists x[\neg P(x)].$$

In some fields, it is common to use infix notation for binary relations and functions, instead of the prefix notation defined above. For example, in arithmetic, one typically writes "2 + 2 = 4" instead of "=(+(2,2),4)". It is common to regard formulas in infix notation as abbreviations for the corresponding formulas in prefix notation, cf. also Term (logic)#Term structure vs. representation.

The definitions above use infix notation for binary connectives such as $\rightarrow$. A less common convention is Polish notation, in which one writes $\rightarrow$, $\wedge$, and so on in front of their arguments rather than between them. This convention allows all punctuation symbols to be discarded. Polish notation is compact and elegant, but rarely used in practice because it is hard for humans to read it. In Polish notation, the formula:

$$\forall x \forall y (P(f(x)) \rightarrow \neg(P(x) \rightarrow Q(f(y), x, z)))$$

becomes "$\forall x \forall y \rightarrow Pfx \neg \rightarrow PxQfyxz$".

Free and Bound Variables

In a formula, a variable may occur free or bound. Intuitively, a variable is free in a formula if it is not quantified: in $\forall y P(x, y)$, variable x is free while y is bound. The free and bound variables of a formula are defined inductively as follows.

1. Atomic formulas. If φ is an atomic formula then x is free in φ if and only if x occurs in φ. Moreover, there are no bound variables in any atomic formula.

2. Negation. x is free in $\neg\varphi$ if and only if x is free in φ. x is bound in $\neg\varphi$ if and only if x is bound in φ.

3. Binary connectives. x is free in $(\varphi \to \psi)$ if and only if x is free in either φ or ψ. x is bound in $(\varphi \to \psi)$ if and only if x is bound in either φ or ψ. The same rule applies to any other binary connective in place of $\to$.

4. Quantifiers. x is free in $\forall y\, \varphi$ if and only if x is free in φ and x is a different symbol from y. Also, x is bound in $\forall y\, \varphi$ if and only if x is y or x is bound in φ. The same rule holds with $\exists$ in place of $\forall$.

For example, in $\forall x\, \forall y\, (P(x) \to Q(x, f(x), z))$, x and y are bound variables, z is a free variable, and w is neither because it does not occur in the formula.

Free and bound variables of a formula need not be disjoint sets: x is both free and bound in $P(x) \to \forall x Q(x)$.

Freeness and boundness can be also specialized to specific occurrences of variables in a formula. For example, in $P(x) \to \forall x Q(x)$, the first occurrence of x is free while the second is bound. In other words, the x in $P(x)$ is free while the x in $\forall x Q(x)$ is bound.

A formula in first-order logic with no free variables is called a first-order sentence. These are the formulas that will have well-defined truth values under an interpretation. For example, whether a formula such as $\text{Phil}(x)$ is true must depend on what x represents. But the sentence $\exists x \text{Phil}(x)$ will be either true or false in a given interpretation.

Example: Ordered Abelian Groups

In mathematics the language of ordered abelian groups has one constant symbol 0, one unary function symbol $-$, one binary function symbol $+$, and one binary relation symbol $\leq$. Then:

*   The expressions $+(x, y)$ and $+(x, +(y, -(z)))$ are terms. These are usually written as $x + y$ and $x + y - z$.

*   The expressions $+(x, y) = 0$ and $\leq(+(x, +(y, -(z))), +(x, y))$ are atomic formulas. These are usually written as $x + y = 0$ and $x + y - z \leq x + y$.

*   The expression $(\forall x \forall y [\leq (+(x, y), z) \to \forall x \forall y + (x, y) = 0)]$ is a formula, which is usually written as $\forall x \forall y (x + y \leq z) \to \forall x \forall y (x + y = 0)$. This formula has one free variable, z.

The axioms for ordered abelian groups can be expressed as a set of sentences in the language. For example, the axiom stating that the group is commutative is usually written:

$$(\forall x)(\forall y)[x + y = y + x].$$

Semantics

An interpretation of a first-order language assigns a denotation to all non-logical constants in that language. It also determines a domain of discourse that specifies the range of the quantifiers. The result is that each term is assigned an object that it represents, and each sentence is assigned a truth value. In this way, an interpretation provides semantic meaning to the terms and formulas of the language. The study of the interpretations of formal languages is called formal semantics. What follows is a description of the standard or Tarskian semantics for first-order logic. (It is also possible to define game semantics for first-order logic, but aside from requiring the axiom of choice, game semantics agree with Tarskian semantics for first-order logic, so game semantics will not be elaborated herein.)

The domain of discourse D is a nonempty set of "objects" of some kind. Intuitively, a first-order formula is a statement about these objects; for example, $\exists x P(x)$ states the existence of an object x such that the predicate P is true where referred to it. The domain of discourse is the set of considered objects. For example, one can take D to be the set of integer numbers.

The interpretation of a function symbol is a function. For example, if the domain of discourse consists of integers, a function symbol f of arity 2 can be interpreted as the function that gives the sum of its arguments. In other words, the symbol f is associated with the function $I(f)$ which, in this interpretation, is addition.

The interpretation of a constant symbol is a function from the one-element set D^0 to D, which can be simply identified with an object in D. For example, an interpretation may assign the value $I(c) = 10$ to the constant symbol c.

The interpretation of an n-ary predicate symbol is a set of n-tuples of elements of the domain of discourse. This means that, given an interpretation, a predicate symbol, and n elements of the domain of discourse, one can tell whether the predicate is true of those elements according to the given interpretation. For example, an interpretation $I(P)$ of a binary predicate symbol P may be the set of pairs of integers such that the first one is less than the second. According to this interpretation, the predicate P would be true if its first argument is less than the second.

First-order Structures

The most common way of specifying an interpretation (especially in mathematics) is to specify a structure (also called a model). The structure consists of a nonempty set D that forms the domain of discourse and an interpretation I of the non-logical terms of the signature. This interpretation is itself a function:

- Each function symbol f of arity n is assigned a function $I(f)$ from D^n to D. In particular, each constant symbol of the signature is assigned an individual in the domain of discourse.

- Each predicate symbol P of arity n is assigned a relation $I(P)$ over D^n or, equivalently, a function from D^n to $\{true, false\}$. Thus each predicate symbol is interpreted by a Boolean-valued function on D.

Evaluation of Truth Values

A formula evaluates to true or false given an interpretation, and a variable assignment μ that associates an element of the domain of discourse with each variable. The reason that a variable assignment is required is to give meanings to formulas with free variables, such as $y = x$. The truth value of this formula changes depending on whether x and y denote the same individual.

First, the variable assignment μ can be extended to all terms of the language, with the result that each term maps to a single element of the domain of discourse. The following rules are used to make this assignment:

1. Variables. Each variable x evaluates to $\mu(x)$

2. Functions. Given terms $t_1,\ldots,t_n$ that have been evaluated to elements $d_1,\ldots,d_n$ of the domain of discourse, and a n-ary function symbol f, the term $f(t_1,\ldots,t_n)$ evaluates to $(I(f))(d_1,\ldots,d_n)$.

Next, each formula is assigned a truth value. The inductive definition used to make this assignment is called the T-schema.

1. Atomic formulas (1). A formula $P(t_1,\ldots,t_n)$ is associated the value true or false depending on whether $\langle v_1,\ldots,v_n \rangle \in I(P)$, where $v_1,\ldots,v_n$ are the evaluation of the terms $t_1,\ldots,t_n$ and $I(P)$ is the interpretation of P, which by assumption is a subset of D^n.

2. Atomic formulas (2). A formula $t_1 = t_2$ is assigned true if t_1 and t_2 evaluate to the same object of the domain of discourse.

3. Logical connectives. A formula in the form $\neg\phi$, $\phi \to \psi$, etc. is evaluated according to the truth table for the connective in question, as in propositional logic.

4. Existential quantifiers. A formula $\exists x\phi(x)$ is true according to M and μ if there exists an evaluation μ' of the variables that only differs from μ regarding the evaluation of x and such that ϕ is true according to the interpretation M and the variable assignment μ'. This formal definition captures the idea that $\exists x\phi(x)$ is true if and only if there is a way to choose a value for x such that $\phi(x)$ is satisfied.

5. Universal quantifiers. A formula $\forall x\phi(x)$ is true according to M and μ if $\phi(x)$ is true for every pair composed by the interpretation M and some variable assignment μ' that differs from μ only on the value of x. This captures the idea that $\forall x\phi(x)$ is true if every possible choice of a value for x causes $\phi(x)$ to be true.

If a formula does not contain free variables, and so is a sentence, then the initial variable assignment does not affect its truth value. In other words, a sentence is true according to M and μ if and only if it is true according to M and every other variable assignment μ'.

There is a second common approach to defining truth values that does not rely on variable assignment functions. Instead, given an interpretation M, one first adds to the signature a collection of constant symbols, one for each element of the domain of discourse in M; say that for each d in the domain the constant symbol c_d is fixed. The interpretation is extended so that each new constant symbol is assigned to its corresponding element of the domain. One now defines truth for quantified formulas syntactically, as follows:

1. Existential quantifiers (alternate). A formula $\exists x\phi(x)$ is true according to M if there is some d in the domain of discourse such that $\phi(c_d)$ holds. Here $\phi(c_d)$ is the result of substituting c_d for every free occurrence of x in φ.

2. Universal quantifiers (alternate). A formula $\forall x\phi(x)$ is true according to M if, for every d in the domain of discourse, $\phi(c_d)$ is true according to M.

This alternate approach gives exactly the same truth values to all sentences as the approach via variable assignments.

Validity, Satisfiability, and Logical Consequence

If a sentence φ evaluates to True under a given interpretation M, one says that M satisfies φ; this is denoted $M \vDash \varphi$. A sentence is satisfiable if there is some interpretation under which it is true.

Satisfiability of formulas with free variables is more complicated, because an interpretation on its own does not determine the truth value of such a formula. The most common convention is that a formula with free variables is said to be satisfied by an interpretation if the formula remains true regardless which individuals from the domain of discourse are assigned to its free variables. This has the same effect as saying that a formula is satisfied if and only if its universal closure is satisfied.

A formula is logically valid (or simply valid) if it is true in every interpretation. These formulas play a role similar to tautologies in propositional logic.

A formula φ is a logical consequence of a formula ψ if every interpretation that makes ψ true also makes φ true. In this case one says that φ is logically implied by ψ.

Algebraizations

An alternate approach to the semantics of first-order logic proceeds via abstract algebra. This approach generalizes the Lindenbaum–Tarski algebras of propositional logic. There are three ways of eliminating quantified variables from first-order logic that do not involve replacing quantifiers with other variable binding term operators:

- Cylindric algebra, by Alfred Tarski and his coworkers;

- Polyadic algebra, by Paul Halmos;

- Predicate functor logic, mainly due to Willard Quine.

These algebras are all lattices that properly extend the two-element Boolean algebra.

Tarski and Givant (1987) showed that the fragment of first-order logic that has no atomic sentence lying in the scope of more than three quantifiers has the same expressive power as relation algebra. This fragment is of great interest because it suffices for Peano arithmetic and most axiomatic set theory, including the canonical ZFC. They also prove that first-order logic with a primitive ordered pair is equivalent to a relation algebra with two ordered pair projection functions.

First-order Theories, Models, and Elementary Classes

A first-order theory of a particular signature is a set of axioms, which are sentences consisting of symbols from that signature. The set of axioms is often finite or recursively enumerable, in which case the theory is called effective. Some authors require theories to also include all logical consequences of the axioms. The axioms are considered to hold within the theory and from them other sentences that hold within the theory can be derived.

A first-order structure that satisfies all sentences in a given theory is said to be a model of the theory. An elementary class is the set of all structures satisfying a particular theory. These classes are a main subject of study in model theory.

Many theories have an intended interpretation, a certain model that is kept in mind when studying the theory. For example, the intended interpretation of Peano arithmetic consists of the usual natural numbers with their usual operations. However, the Löwenheim–Skolem theorem shows that most first-order theories will also have other, nonstandard models.

A theory is consistent if it is not possible to prove a contradiction from the axioms of the theory. A theory is complete if, for every formula in its signature, either that formula or its negation is a logical consequence of the axioms of the theory. Gödel's incompleteness theorem shows that effective first-order theories that include a sufficient portion of the theory of the natural numbers can never be both consistent and complete.

Empty Domains

The definition above requires that the domain of discourse of any interpretation must be a nonempty set. There are settings, such as inclusive logic, where empty domains are permitted. Moreover, if a class of algebraic structures includes an empty structure (for example, there is an empty poset), that class can only be an elementary class in first-order logic if empty domains are permitted or the empty structure is removed from the class.

There are several difficulties with empty domains, however:

- Many common rules of inference are only valid when the domain of discourse is required to be nonempty. One example is the rule stating that $\phi \vee \exists x \psi$ implies $\exists x(\phi \vee \psi)$ when x is not a free variable in ϕ. This rule, which is used to put formulas into prenex normal form, is sound in nonempty domains, but unsound if the empty domain is permitted.

- The definition of truth in an interpretation that uses a variable assignment function cannot work with empty domains, because there are no variable assignment functions whose range is empty. (Similarly, one cannot assign interpretations to constant symbols.) This truth definition requires that one must select a variable assignment function (μ above) before truth values for even atomic formulas can be defined. Then the truth value of a sentence is defined to be its truth value under any variable assignment, and it is proved that this truth value does not depend on which assignment is chosen. This technique does not work if there are no assignment functions at all; it must be changed to accommodate empty domains.

Thus, when the empty domain is permitted, it must often be treated as a special case. Most authors, however, simply exclude the empty domain by definition.

Deductive Systems

A deductive system is used to demonstrate, on a purely syntactic basis, that one formula is a logical consequence of another formula. There are many such systems for first-order logic, including Hilbert-style deductive systems, natural deduction, the sequent calculus, the tableaux method, and resolution. These share the common property that a deduction is a finite syntactic object; the format of this object, and the way it is constructed, vary widely. These finite deductions themselves are often called derivations in proof theory. They are also often called proofs, but are completely formalized unlike natural-language mathematical proofs.

A deductive system is sound if any formula that can be derived in the system is logically valid. Conversely, a deductive system is complete if every logically valid formula is derivable. All of the systems discussed in this article are both sound and complete. They also share the property that it is possible to effectively verify that a purportedly valid deduction is actually a deduction; such deduction systems are called effective.

A key property of deductive systems is that they are purely syntactic, so that derivations can be verified without considering any interpretation. Thus a sound argument is correct in every possible interpretation of the language, regardless whether that interpretation is about mathematics, economics, or some other area.

In general, logical consequence in first-order logic is only semidecidable: if a sentence A logically implies a sentence B then this can be discovered (for example, by searching for a proof until one is found, using some effective, sound, complete proof system). However, if A does not logically imply B, this does not mean that A logically implies the negation of B. There is no effective procedure that, given formulas A and B, always correctly decides whether A logically implies B.

Rules of Inference

A rule of inference states that, given a particular formula (or set of formulas) with a certain property as a hypothesis, another specific formula (or set of formulas) can be derived as a conclusion. The rule is sound (or truth-preserving) if it preserves validity in the sense that whenever any interpretation satisfies the hypothesis, that interpretation also satisfies the conclusion.

For example, one common rule of inference is the rule of substitution. If t is a term and φ is a formula possibly containing the variable x, then $\varphi[t/x]$ is the result of replacing all free instances of x by t in φ. The substitution rule states that for any φ and any term t, one can conclude $\varphi[t/x]$

from φ provided that no free variable of t becomes bound during the substitution process. (If some free variable of t becomes bound, then to substitute t for x it is first necessary to change the bound variables of φ to differ from the free variables of t.)

To see why the restriction on bound variables is necessary, consider the logically valid formula φ given by $\exists x(x = y)$, in the signature of $(0,1,+,\times,=)$ of arithmetic. If t is the term "x + 1", the formula $\varphi[t/y]$ is $\exists x(x = x + 1)$, which will be false in many interpretations. The problem is that the free variable x of t became bound during the substitution. The intended replacement can be obtained by renaming the bound variable x of φ to something else, say z, so that the formula after substitution is $\exists z(z = x + 1)$, which is again logically valid.

The substitution rule demonstrates several common aspects of rules of inference. It is entirely syntactical; one can tell whether it was correctly applied without appeal to any interpretation. It has (syntactically defined) limitations on when it can be applied, which must be respected to preserve the correctness of derivations. Moreover, as is often the case, these limitations are necessary because of interactions between free and bound variables that occur during syntactic manipulations of the formulas involved in the inference rule.

Hilbert-style Systems and Natural Deduction

A deduction in a Hilbert-style deductive system is a list of formulas, each of which is a logical axiom, a hypothesis that has been assumed for the derivation at hand, or follows from previous formulas via a rule of inference. The logical axioms consist of several axiom schemas of logically valid formulas; these encompass a significant amount of propositional logic. The rules of inference enable the manipulation of quantifiers. Typical Hilbert-style systems have a small number of rules of inference, along with several infinite schemas of logical axioms. It is common to have only modus ponens and universal generalization as rules of inference.

Natural deduction systems resemble Hilbert-style systems in that a deduction is a finite list of formulas. However, natural deduction systems have no logical axioms; they compensate by adding additional rules of inference that can be used to manipulate the logical connectives in formulas in the proof.

Sequent Calculus

The sequent calculus was developed to study the properties of natural deduction systems. Instead of working with one formula at a time, it uses sequents, which are expressions of the form

$$A_1, \ldots, A_n \vdash B_1, \ldots, B_k,$$

where $A_1, \ldots, A_n, B_1, \ldots, B_k$ are formulas and the turnstile symbol $\vdash$ is used as punctuation to separate the two halves. Intuitively, a sequent expresses the idea that $(A_1 \wedge \cdots \wedge A_n)$ implies $(B_1 \vee \cdots \vee B_k)$.

Tableaux Method

Unlike the methods just described, the derivations in the tableaux method are not lists of formulas. Instead, a derivation is a tree of formulas. To show that a formula A is provable, the tableaux method attempts to demonstrate that the negation of A is unsatisfiable. The tree of the derivation has $\neg A$ at its root; the tree branches in a way that reflects the structure of the formula. For example, to show that $C \vee D$ is unsatisfiable requires showing that C and D are each unsatisfiable; this corresponds to a branching point in the tree with parent $C \vee D$ and children C and D.

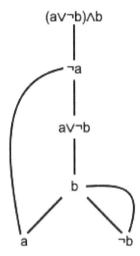

A tableaux proof for the propositional formula $((a \vee \neg b) \wedge b) \rightarrow a$.

Resolution

The resolution rule is a single rule of inference that, together with unification, is sound and complete for first-order logic. As with the tableaux method, a formula is proved by showing that the negation of the formula is unsatisfiable. Resolution is commonly used in automated theorem proving.

The resolution method works only with formulas that are disjunctions of atomic formulas; arbitrary formulas must first be converted to this form through Skolemization. The resolution rule states that from the hypotheses $A_1 \vee \cdots \vee A_k \vee C$ and $B_1 \vee \cdots \vee B_l \vee \neg C$, the conclusion $A_1 \vee \cdots \vee A_k \vee B_1 \vee \cdots \vee B_l$ can be obtained.

Provable Identities

Many identities can be proved, which establish equivalences between particular formulas. These identities allow for rearranging formulas by moving quantifiers across other connectives, and are useful for putting formulas in prenex normal form. Some provable identities include:

$$\neg \forall x P(x) \Leftrightarrow \exists x \neg P(x)$$

$$\neg \exists x P(x) \Leftrightarrow \forall x \neg P(x)$$

$$\forall x \forall y P(x, y) \Leftrightarrow \forall y \forall x P(x, y)$$

$$\exists x \exists y P(x, y) \Leftrightarrow \exists y \exists x P(x, y)$$

$$\forall x P(x) \wedge \forall x Q(x) \Leftrightarrow \forall x (P(x) \wedge Q(x))$$

$$\exists x P(x) \vee \exists x Q(x) \Leftrightarrow \exists x (P(x) \vee Q(x))$$

$$P \wedge \exists x Q(x) \Leftrightarrow \exists x (P \wedge Q(x)) \text{ (where } x \text{ must not occur free in } P)$$

$$P \vee \forall x Q(x) \Leftrightarrow \forall x (P \vee Q(x)) \text{ (where } x \text{ must not occur free in } P)$$

Equality and its Axioms

There are several different conventions for using equality (or identity) in first-order logic. The most common convention, known as first-order logic with equality, includes the equality symbol as a primitive logical symbol which is always interpreted as the real equality relation between members of the domain of discourse, such that the "two" given members are the same member. This approach also adds certain axioms about equality to the deductive system employed. These equality axioms are:

1. Reflexivity. For each variable x, $x = x$.

2. Substitution for functions. For all variables x and y, and any function symbol f,

    $$x = y \rightarrow f(...,x,...) = f(...,y,...).$$

3. Substitution for formulas. For any variables x and y and any formula $\varphi(x)$, if φ' is obtained by replacing any number of free occurrences of x in φ with y, such that these remain free occurrences of y, then

    $$x = y \rightarrow (\varphi \rightarrow \varphi').$$

These are axiom schemas, each of which specifies an infinite set of axioms. The third schema is known as Leibniz's law, "the principle of substitutivity", "the indiscernibility of identicals", or "the replacement property". The second schema, involving the function symbol f, is (equivalent to) a special case of the third schema, using the formula

$$x = y \rightarrow (f(...,x,...) = z \rightarrow f(...,y,...) = z).$$

Many other properties of equality are consequences of the axioms above, for example:

1. Symmetry. If $x = y$ then $y = x$.

2. Transitivity. If $x = y$ and $y = z$ then $x = z$.

First-order Logic without Equality

An alternate approach considers the equality relation to be a non-logical symbol. This convention is known as first-order logic without equality. If an equality relation is included in the signature, the axioms of equality must now be added to the theories under consideration, if desired, instead of being considered rules of logic. The main difference between this method and first-order logic with equality is that an interpretation may now interpret two distinct individuals as "equal" (although, by Leibniz's law, these will satisfy exactly the same formulas under any interpretation). That is, the equality relation may now be interpreted by an arbitrary equivalence relation on the domain of discourse that is congruent with respect to the functions and relations of the interpretation.

When this second convention is followed, the term normal model is used to refer to an interpretation where no distinct individuals a and b satisfy $a = b$. In first-order logic with equality, only normal models are considered, and so there is no term for a model other than a normal model. When first-order logic without equality is studied, it is necessary to amend the statements of results such as the Löwenheim–Skolem theorem so that only normal models are considered.

First-order logic without equality is often employed in the context of second-order arithmetic and other higher-order theories of arithmetic, where the equality relation between sets of natural numbers is usually omitted.

Defining Equality within a Theory

If a theory has a binary formula $A(x,y)$ which satisfies reflexivity and Leibniz's law, the theory is said to have equality, or to be a theory with equality. The theory may not have all instances of the above schemas as axioms, but rather as derivable theorems. For example, in theories with no function symbols and a finite number of relations, it is possible to define equality in terms of the relations, by defining the two terms s and t to be equal if any relation is unchanged by changing s to t in any argument.

Some theories allow other *ad hoc* definitions of equality:

- In the theory of partial orders with one relation symbol $\leq$, one could define $s = t$ to be an abbreviation for $s \leq t \wedge t \leq s$.

- In set theory with one relation $\in$, one may define $s = t$ to be an abbreviation for $\forall x(s \in x \leftrightarrow t \in x) \wedge \forall x(x \in s \leftrightarrow x \in t)$. This definition of equality then automatically satisfies the axioms for equality. In this case, one should replace the usual axiom of extensionality, which can be stated as $\forall x \forall y[\forall z(z \in x \Leftrightarrow z \in y) \Rightarrow x = y]$, with an alternative formulation $\forall x \forall y[\forall z(z \in x \Leftrightarrow z \in y) \Rightarrow \forall z(x \in z \Leftrightarrow y \in z)]$, which says that if sets x and y have the same elements, then they also belong to the same sets.

Metalogical Properties

One motivation for the use of first-order logic, rather than higher-order logic, is that first-order logic has many metalogical properties that stronger logics do not have. These results concern general properties of first-order logic itself, rather than properties of individual theories. They provide fundamental tools for the construction of models of first-order theories.

Completeness and Undecidability

Gödel's completeness theorem, proved by Kurt Gödel in 1929, establishes that there are sound, complete, effective deductive systems for first-order logic, and thus the first-order logical consequence relation is captured by finite provability. Naively, the statement that a formula φ logically implies a formula ψ depends on every model of φ; these models will in general be of arbitrarily large cardinality, and so logical consequence cannot be effectively verified by checking every model. However, it is possible to enumerate all finite derivations and search for a derivation of ψ from φ. If ψ is logically implied by φ, such a derivation will eventually be found. Thus first-order logical consequence is semidecidable: it is possible to make an effective enumeration of all pairs of sentences (φ, ψ) such that ψ is a logical consequence of φ.

Unlike propositional logic, first-order logic is undecidable (although semidecidable), provided that the language has at least one predicate of arity at least 2 (other than equality). This means that there is no decision procedure that determines whether arbitrary formulas are logically valid. This result was established independently by Alonzo Church and Alan Turing in 1936 and 1937, respectively, giving a negative answer to the Entscheidungsproblem posed by David Hilbert in 1928. Their proofs demonstrate a connection between the unsolvability of the decision problem for first-order logic and the unsolvability of the halting problem.

There are systems weaker than full first-order logic for which the logical consequence relation is decidable. These include propositional logic and monadic predicate logic, which is first-order logic restricted to unary predicate symbols and no function symbols. Other logics with no function symbols which are decidable are the guarded fragment of first-order logic, as well as two-variable logic. The Bernays–Schönfinkel class of first-order formulas is also decidable. Decidable subsets of first-order logic are also studied in the framework of description logics.

The Löwenheim–Skolem Theorem

The Löwenheim–Skolem theorem shows that if a first-order theory of cardinality λ has an infinite model, then it has models of every infinite cardinality greater than or equal to λ. One of the earliest results in model theory, it implies that it is not possible to characterize countability or uncountability in a first-order language. That is, there is no first-order formula $\varphi(x)$ such that an arbitrary structure M satisfies φ if and only if the domain of discourse of M is countable (or, in the second case, uncountable).

The Löwenheim–Skolem theorem implies that infinite structures cannot be categorically axiomatized in first-order logic. For example, there is no first-order theory whose only model is the real line: any first-order theory with an infinite model also has a model of cardinality larger than the continuum. Since the real line is infinite, any theory satisfied by the real line is also satisfied by some nonstandard models. When the Löwenheim–Skolem theorem is applied to first-order set theories, the nonintuitive consequences are known as Skolem's paradox.

The Compactness Theorem

The compactness theorem states that a set of first-order sentences has a model if and only if every finite subset of it has a model. This implies that if a formula is a logical consequence of an infinite set of first-order axioms, then it is a logical consequence of some finite number of those axioms. This theorem was proved first by Kurt Gödel as a consequence of the completeness theorem, but many additional proofs have been obtained over time. It is a central tool in model theory, providing a fundamental method for constructing models.

The compactness theorem has a limiting effect on which collections of first-order structures are elementary classes. For example, the compactness theorem implies that any theory that has arbitrarily large finite models has an infinite model. Thus the class of all finite graphs is not an elementary class (the same holds for many other algebraic structures).

There are also more subtle limitations of first-order logic that are implied by the compactness theorem. For example, in computer science, many situations can be modeled as a directed graph of states (nodes) and connections (directed edges). Validating such a system may require showing that no "bad" state can be reached from any "good" state. Thus one seeks to determine if the good and bad states are in different connected components of the graph. However, the compactness theorem can be used to show that connected graphs are not an elementary class in first-order logic, and there is no formula $\varphi(x,y)$ of first-order logic, in the logic of graphs, that expresses the idea that there is a path from x to y. Connectedness can be expressed in second-order logic, however, but not with only existential set quantifiers, as Σ_1^1 also enjoys compactness.

Lindström's Theorem

Per Lindström showed that the metalogical properties just discussed actually characterize first-order logic in the sense that no stronger logic can also have those properties (Ebbinghaus and Flum 1994). Lindström defined a class of abstract logical systems, and a rigorous definition of the relative strength of a member of this class. He established two theorems for systems of this type:

- A logical system satisfying Lindström's definition that contains first-order logic and satisfies both the Löwenheim–Skolem theorem and the compactness theorem must be equivalent to first-order logic.

- A logical system satisfying Lindström's definition that has a semidecidable logical consequence relation and satisfies the Löwenheim–Skolem theorem must be equivalent to first-order logic.

Limitations

Although first-order logic is sufficient for formalizing much of mathematics, and is commonly used in computer science and other fields, it has certain limitations. These include limitations on its expressiveness and limitations of the fragments of natural languages that it can describe.

For instance, first-order logic is undecidable, meaning a sound, complete and terminating decision algorithm is impossible. This has led to the study of interesting decidable fragments such as C_2, first-order logic with two variables and the counting quantifiers $\exists^{\geq n}$ and $\exists^{\leq n}$ (these quantifiers are, respectively, "there exists at least n" and "there exists at most n") (Horrocks 2010).

Expressiveness

The Löwenheim–Skolem theorem shows that if a first-order theory has any infinite model, then it has infinite models of every cardinality. In particular, no first-order theory with an infinite model can be categorical. Thus there is no first-order theory whose only model has the set of natural numbers as its domain, or whose only model has the set of real numbers as its domain. Many extensions of first-order logic, including infinitary logics and higher-order logics, are more expressive in the sense that they do permit categorical axiomatizations of the natural numbers or real numbers. This expressiveness comes at a metalogical cost, however: by Lindström's theorem, the compactness theorem and the downward Löwenheim–Skolem theorem cannot hold in any logic stronger than first-order.

Formalizing Natural Languages

First-order logic is able to formalize many simple quantifier constructions in natural language, such as "every person who lives in Perth lives in Australia". But there are many more complicated features of natural language that cannot be expressed in (single-sorted) first-order logic. "Any logical system which is appropriate as an instrument for the analysis of natural language needs a much richer structure than first-order predicate logic" (Gamut 1991, p. 75).

Type	Example	Comment
Quantification over properties	If John is self-satisfied, then there is at least one thing he has in common with Peter	Requires a quantifier over predicates, which cannot be implemented in single-sorted first-order logic: $Zj \rightarrow \exists X(Xj \wedge Xp)$
Quantification over properties	Santa Claus has all the attributes of a sadist	Requires quantifiers over predicates, which cannot be implemented in single-sorted first-order logic: $\forall X(\forall x(Sx \rightarrow Xx) \rightarrow Xs)$
Predicate adverbial	John is walking quickly	Cannot be analysed as $Wj \wedge Qj$; predicate adverbials are not the same kind of thing as second-order predicates such as colour
Relative adjective	Jumbo is a small elephant	Cannot be analysed as $Sj \wedge Ej$; predicate adjectives are not the same kind of thing as second-order predicates such as colour
Predicate adverbial modifier	John is walking very quickly	-
Relative adjective modifier	Jumbo is terribly small	An expression such as "terribly", when applied to a relative adjective such as "small", results in a new composite relative adjective "terribly small"
Prepositions	Mary is sitting next to John	The preposition "next to" when applied to "John" results in the predicate adverbial "next to John"

Restrictions, Extensions, and Variations

There are many variations of first-order logic. Some of these are inessential in the sense that they merely change notation without affecting the semantics. Others change the expressive power more significantly, by extending the semantics through additional quantifiers or other new logical symbols. For example, infinitary logics permit formulas of infinite size, and modal logics add symbols for possibility and necessity.

Restricted Languages

First-order logic can be studied in languages with fewer logical symbols than were described above.

- Because $\exists x\phi(x)$ can be expressed as $\neg\forall x\neg\phi(x)$, and $\forall x\phi(x)$ can be expressed as $\neg\exists x\neg\phi(x)$, either of the two quantifiers $\exists$ and $\forall$ can be dropped.

- Since $\phi \vee \emptyset$ can be expressed as $\neg(\neg\phi \wedge \neg\emptyset)$ and $\phi \wedge \emptyset$ can be expressed as $\neg(\neg\phi \vee \neg\emptyset)$, either $\vee$ or $\wedge$ can be dropped. In other words, it is sufficient to have $\neg$ and $\wedge$, or $\neg$ and $\vee$, as the only logical connectives.

- Similarly, it is sufficient to have only $\neg$ and $\rightarrow$ as logical connectives, or to have only the Sheffer stroke (NAND) or the Peirce arrow (NOR) operator.

- It is possible to entirely avoid function symbols and constant symbols, rewriting them via predicate symbols in an appropriate way. For example, instead of using a constant symbol 0 one may use a predicate $0(x)$ (interpreted as $x = 0$), and replace every predicate such as $P(0, y)$ with $\forall x\, (0(x) \to P(x, y))$. A function such as $f(x_1, x_2, ..., x_n)$ will similarly be replaced by a predicate $F(x_1, x_2, ..., x_n, y)$ interpreted as $y = f(x_1, x_2, ..., x_n)$. This change requires adding additional axioms to the theory at hand, so that interpretations of the predicate symbols used have the correct semantics.

Restrictions such as these are useful as a technique to reduce the number of inference rules or axiom schemas in deductive systems, which leads to shorter proofs of metalogical results. The cost of the restrictions is that it becomes more difficult to express natural-language statements in the formal system at hand, because the logical connectives used in the natural language statements must be replaced by their (longer) definitions in terms of the restricted collection of logical connectives. Similarly, derivations in the limited systems may be longer than derivations in systems that include additional connectives. There is thus a trade-off between the ease of working within the formal system and the ease of proving results about the formal system.

It is also possible to restrict the arities of function symbols and predicate symbols, in sufficiently expressive theories. One can in principle dispense entirely with functions of arity greater than 2 and predicates of arity greater than 1 in theories that include a pairing function. This is a function of arity 2 that takes pairs of elements of the domain and returns an ordered pair containing them. It is also sufficient to have two predicate symbols of arity 2 that define projection functions from an ordered pair to its components. In either case it is necessary that the natural axioms for a pairing function and its projections are satisfied.

Many-sorted logic

Ordinary first-order interpretations have a single domain of discourse over which all quantifiers range. Many-sorted first-order logic allows variables to have different sorts, which have different domains. This is also called typed first-order logic, and the sorts called types (as in data type), but it is not the same as first-order type theory. Many-sorted first-order logic is often used in the study of second-order arithmetic.

When there are only finitely many sorts in a theory, many-sorted first-order logic can be reduced to single-sorted first-order logic. One introduces into the single-sorted theory a unary predicate symbol for each sort in the many-sorted theory, and adds an axiom saying that these unary predicates partition the domain of discourse. For example, if there are two sorts, one adds predicate symbols $P_1(x)$ and $P_2(x)$ and the axiom

$$\forall x (P_1(x) \lor P_2(x)) \land \neg \exists x (P_1(x) \land P_2(x)).$$

Then the elements satisfying P_1 are thought of as elements of the first sort, and elements satisfying P_2 as elements of the second sort. One can quantify over each sort by using the corresponding predicate symbol to limit the range of quantification. For example, to say there is an element of the first sort satisfying formula $\varphi(x)$, one writes

$$\exists x (P_1(x) \land \phi(x)).$$

Additional Quantifiers

Additional quantifiers can be added to first-order logic.

- Sometimes it is useful to say that "$P(x)$ holds for exactly one x", which can be expressed as $\exists! x\, P(x)$. This notation, called uniqueness quantification, may be taken to abbreviate a formula such as $\exists x\, (P(x) \wedge \forall y\, (P(y) \to (x = y)))$.

- First-order logic with extra quantifiers has new quantifiers $Qx,...,$ with meanings such as "there are many x such that ...". Bounded quantifiers are often used in the study of set theory or arithmetic.

Infinitary Logics

Infinitary logic allows infinitely long sentences. For example, one may allow a conjunction or disjunction of infinitely many formulas, or quantification over infinitely many variables. Infinitely long sentences arise in areas of mathematics including topology and model theory.

Infinitary logic generalizes first-order logic to allow formulas of infinite length. The most common way in which formulas can become infinite is through infinite conjunctions and disjunctions. However, it is also possible to admit generalized signatures in which function and relation symbols are allowed to have infinite arities, or in which quantifiers can bind infinitely many variables. Because an infinite formula cannot be represented by a finite string, it is necessary to choose some other representation of formulas; the usual representation in this context is a tree. Thus formulas are, essentially, identified with their parse trees, rather than with the strings being parsed.

The most commonly studied infinitary logics are denoted $L_{\alpha\beta}$, where α and β are each either cardinal numbers or the symbol ∞. In this notation, ordinary first-order logic is $L_{\omega\omega}$. In the logic $L_{\infty\omega}$, arbitrary conjunctions or disjunctions are allowed when building formulas, and there is an unlimited supply of variables. More generally, the logic that permits conjunctions or disjunctions with less than κ constituents is known as $L_{\kappa\omega}$. For example, $L_{\omega_1\omega}$ permits countable conjunctions and disjunctions.

The set of free variables in a formula of $L_{\kappa\omega}$ can have any cardinality strictly less than κ, yet only finitely many of them can be in the scope of any quantifier when a formula appears as a subformula of another. In other infinitary logics, a subformula may be in the scope of infinitely many quantifiers. For example, in $L_{\kappa\infty}$, a single universal or existential quantifier may bind arbitrarily many variables simultaneously. Similarly, the logic $L_{\kappa\lambda}$ permits simultaneous quantification over fewer than λ variables, as well as conjunctions and disjunctions of size less than κ.

Non-classical and Modal Logics

- Intuitionistic first-order logic uses intuitionistic rather than classical propositional calculus; for example, $\neg\neg\varphi$ need not be equivalent to φ.

- First-order modal logic allows one to describe other possible worlds as well as this contingently true world which we inhabit. In some versions, the set of possible worlds varies depending on which possible world one inhabits. Modal logic has extra *modal operators* with meanings which can be characterized informally as, for example "it is necessary that φ" (true in all possible worlds) and "it is possible that φ" (true in some possible world). With standard first-order logic we have a single domain and each predicate is assigned one extension. With first-order modal logic we have a *domain function* that assigns each possible world its own domain, so that each predicate gets an extension only relative to these possible worlds. This allows us to model cases where, for example, Alex is a Philosopher, but might have been a Mathematician, and might not have existed at all. In the first possible world $P(a)$ is true, in the second $P(a)$ is false, and in the third possible world there is no a in the domain at all.

- first-order fuzzy logics are first-order extensions of propositional fuzzy logics rather than classical propositional calculus.

Fixpoint Logic

Fixpoint logic extends first-order logic by adding the closure under the least fixed points of positive operators.

Higher-order Logics

The characteristic feature of first-order logic is that individuals can be quantified, but not predicates. Thus

$$\exists a(\text{Phil}(a))$$

is a legal first-order formula, but

$$\exists \text{Phil}(\text{Phil}(a))$$

is not, in most formalizations of first-order logic. Second-order logic extends first-order logic by adding the latter type of quantification. Other higher-order logics allow quantification over even higher types than second-order logic permits. These higher types include relations between relations, functions from relations to relations between relations, and other higher-type objects. Thus the "first" in first-order logic describes the type of objects that can be quantified.

Unlike first-order logic, for which only one semantics is studied, there are several possible semantics for second-order logic. The most commonly employed semantics for second-order and higher-order logic is known as full semantics. The combination of additional quantifiers and the full semantics for these quantifiers makes higher-order logic stronger than first-order logic. In particular, the (semantic) logical consequence relation for second-order and higher-order logic is not semidecidable; there is no effective deduction system for second-order logic that is sound and complete under full semantics.

Second-order logic with full semantics is more expressive than first-order logic. For example, it is possible to create axiom systems in second-order logic that uniquely characterize the natural numbers and the real line. The cost of this expressiveness is that second-order and higher-order logics have fewer attractive metalogical properties than first-order logic. For example, the Löwenheim–

Skolem theorem and compactness theorem of first-order logic become false when generalized to higher-order logics with full semantics.

Automated Theorem Proving and Formal Methods

Automated theorem proving refers to the development of computer programs that search and find derivations (formal proofs) of mathematical theorems. Finding derivations is a difficult task because the search space can be very large; an exhaustive search of every possible derivation is theoretically possible but computationally infeasible for many systems of interest in mathematics. Thus complicated heuristic functions are developed to attempt to find a derivation in less time than a blind search.

The related area of automated proof verification uses computer programs to check that human-created proofs are correct. Unlike complicated automated theorem provers, verification systems may be small enough that their correctness can be checked both by hand and through automated software verification. This validation of the proof verifier is needed to give confidence that any derivation labeled as "correct" is actually correct.

Some proof verifiers, such as Metamath, insist on having a complete derivation as input. Others, such as Mizar and Isabelle, take a well-formatted proof sketch (which may still be very long and detailed) and fill in the missing pieces by doing simple proof searches or applying known decision procedures: the resulting derivation is then verified by a small, core "kernel". Many such systems are primarily intended for interactive use by human mathematicians: these are known as proof assistants. They may also use formal logics that are stronger than first-order logic, such as type theory. Because a full derivation of any nontrivial result in a first-order deductive system will be extremely long for a human to write, results are often formalized as a series of lemmas, for which derivations can be constructed separately.

Automated theorem provers are also used to implement formal verification in computer science. In this setting, theorem provers are used to verify the correctness of programs and of hardware such as processors with respect to a formal specification. Because such analysis is time-consuming and thus expensive, it is usually reserved for projects in which a malfunction would have grave human or financial consequences.

Ways to Define FOL

Syntax

Let us first introduce the symbols, or alphabet, being used. Beware that there are all sorts of slightly different ways to define FOL.

Alphabet

- Logical Symbols: These are symbols that have a standard meaning, like: AND, OR, NOT, ALL, EXISTS, IMPLIES, IFF, FALSE, =.

- Non-Logical Symbols: divided in:

    o Constants:

- Predicates: 1-ary, 2-ary, .., n-ary. These are usually just identifiers.

- Functions: 0-ary, 1-ary, 2-ary, .., n-ary. These are usually just identifiers. 0-ary functions are also called individual constants.

Where predicates return true or false, functions can return any value.

o Variables: Usually an identifier.

One needs to be able to distinguish the identifiers used for predicates, functions, and variables by using some appropriate convention, for example, capitals for function and predicate symbols and lower cases for variables.

Terms

A Term is either an individual constant (a 0-ary function), or a variable, or an n-ary function applied to n terms: $F(t1\ t2\ ..tn)$

[We will use both the notation $F(t1\ t2\ ..tn)$ and the notation $(F\ t1\ t2\ ..\ tn)$]

Atomic Formulae

An Atomic Formula is either FALSE or an n-ary predicate applied to n terms: $P(t1\ t2\ ..\ tn)$. In the case that "=" is a logical symbol in the language, $(t1 = t2)$, where t1 and t2 are terms, is an atomic formula.

Literals

A Literal is either an atomic formula (a Positive Literal), or the negation of an atomic formula (a Negative Literal). A Ground Literal is a variable-free literal.

Clauses

A Clause is a disjunction of literals. A Ground Clause is a variable-free clause. A Horn Clause is a clause with at most one positive literal. A Definite Clause is a Horn Clause with exactly one positive Literal.

Notice that implications are equivalent to Horn or Definite clauses:

(A IMPLIES B) is equivalent to ((NOT A) OR B)

(A AND B IMPLIES FALSE) is equivalent to ((NOT A) OR (NOT B)).

Formulae

A Formula is either:

- an atomic formula, or

- a Negation, i.e. the NOT of a formula, or

- a Conjunctive Formula, i.e. the AND of formulae, or

- a Disjunctive Formula, i.e. the OR of formulae, or

- an Implication, that is a formula of the form (formula1 IMPLIES formula2), or

- an Equivalence, that is a formula of the form (formula1 IFF formula2), or

- a Universally Quantified Formula, that is a formula of the form (ALL variable formula). We say that occurrences of variable are bound in formula [we should be more precise]. Or

- a Existentially Quantified Formula, that is a formula of the form (EXISTS variable formula). We say that occurrences of variable are bound in formula [we should be more precise].

An occurrence of a variable in a formula that is not bound, is said to be free. A formula where all occurrences of variables are bound is called a closed formula, one where all variables are free is called an open formula.

A formula that is the disjunction of clauses is said to be in Clausal Form. We shall see that there is a sense in which every formula is equivalent to a clausal form.

Often it is convenient to refer to terms and formulae with a single name. Form or Expression is used to this end.

Substitutions

- Given a term s, the result [substitution instance] of substituting a term t in s for a variable x, $s[t/x]$, is:

    o t, if s is the variable x

    o y, if s is the variable y different from x

    o $F(s_1[t/x]\ s_2[t/x] .. s_n[t/x])$, if s is $F(s_1\ s_2 .. s_n)$.

- Given a formula A, the result (substitution instance) of substituting a term t in A for a variable x, $A[t/x]$, is:

    o FALSE, if A is FALSE,

    o $P(t_1[t/x]\ t_2[t/x] .. t_n[t/x])$, if A is $P(t_1\ t_2 .. t_n)$,

    o $(B[t/x]$ AND $C[t/x])$ if A is (B AND C), and similarly for the other connectives,

    o (ALL x B) if A is (ALL x B), (similarly for EXISTS),

    o $(ALL\ y\ B[t/x])$, if A is (ALL y B) and y is different from x (similarly for EXISTS).

The substitution $[t/x]$ can be seen as a map from terms to terms and from formulae to formulae. We can define similarly $[t_1/x_1\ t_2/x_2 .. t_n/x_n]$, where $t_1\ t_2 .. t_n$ are terms and $x_1\ x_2 .. x_n$ are variables, as a map, the [simultaneous] substitution of x_1 by t_1, x_2 by t_2, .., of x_n by t_n. [If all the terms $t_1 .. t_n$ are variables, the substitution is called an alphabetic variant, and if they are ground terms, it is called a ground substitution.] Note that a simultaneous substitution is not the same as a sequential substitution.

Unification

- Given two substitutions S = [t1/x1 .. tn/xn] and V = [u1/y1 .. um/ym], the composition of S and V, S . V, is the substitution obtained by:

    o Applying V to t1 .. tn [the operation on substitutions with just this property is called concatenation], and

    o adding any pair uj/yj such that yj is not in {x1 .. xn}.

    For example: [G(x y)/z].[A/x B/y C/w D/z] is [G(A B)/z A/x B/y C/w].

    Composition is an operation that is associative and non commutative

- A set of forms f1 .. fn is unifiable iff there is a substitution S such that f1.S = f2.S = .. = fn.S. We then say that S is a unifier of the set. For example {P(x F(y) B) P(x F(B) B)} is unified by [A/x B/y] and also unified by [B/y].

- A Most General Unifier (MGU) of a set of forms f1 .. fn is a substitution S that unifies this set and such that for any other substitution T that unifies the set there is a substitution V such that S.V = T. The result of applying the MGU to the forms is called a Most General Instance (MGI). Here are some examples:

FORMULAE	MGU	MGI
(P x), (P A)	[A/x]	(P A)
(P (F x) y (G y)), (P (F x) z (G x))	[x/y x/z]	(P (F x) x (G x))
(F x (G y)), (F (G u) (G z))	[(G u)/x y/z]	(F (G u) (G y))
(F x (G y)), (F (G u) (H z))	Not Unifiable	
(F x (G x) x), (F (G u) (G (G z)) z)	Not Unifiable	

This last example is interesting: we first find that (G u) should replace x, then that (G z) should replace x; finally that x and z are equivalent. So we need x->(G z) and x->z to be both true. This would be possible only if z and (G z) were equivalent. That cannot happen for a finite term. To recognize cases such as this that do not allow unification [we cannot replace z by (G z) since z occurs in (G z)], we need what is called an Occur Test . Most Prolog implementation use Unification extensively but do not do the occur test for efficiency reasons.

The determination of Most General Unifier is done by the Unification Algorithm. Here is the pseudo code for it:

FUNCTION Unify WITH PARAMETERS form1, form2, and assign RETURNS MGU, where form1 and form2 are the forms that we want to unify, and assign is initially nil.

1. Use the Find-Difference function described below to determine the first elements where form1 and form2 differ and one of the elements is a variable. Call difference-set the value returned by Find-Difference. This value will be either the atom Fail, if the two forms cannot be unified; or null, if the two forms are identical; or a pair of the form (Variable Expression).

2. If Find-Difference returned the atom Fail, Unify also returns Fail and we cannot unify the two forms.

3. If Find-Difference returned nil, then Unify will return assign as MGU.

4. Otherwise, we replace each occurrence of Variable by Expression in form1 and form2; we compose the given assignment assign with the assignment that maps Variable into Expression, and we repeat the process for the new form1, form2, and assign.

FUNCTION Find-Difference WITH PARAMETERS form1 and form2 RETURNS pair, where form1 and form2 are e-expressions.

1. If form1 and form2 are the same variable, return nil.

2. Otherwise, if either form1 or form2 is a variable, and it does not appear anywhere in the other form, then return the pair (Variable Other-Form), otherwise return Fail.

3. Otherwise, if either form1 or form2 is an atom then if they are the same atom then return nil otherwise return Fail.

4. Otherwise both form1 and form2 are lists.

   Apply the Find-Difference function to corresponding elements of the two lists until either a call returns a non-null value or the two lists are simultaneously exhausted, or some elements are left over in one list.

   In the first case, that non-null value is returned; in the second, nil is returned; in the third, Fail is returned

Semantics

Before we can continue in the "syntactic" domain with concepts like Inference Rules and Proofs, we need to clarify the Semantics, or meaning, of First Order Logic.

An L-Structure or Conceptualization for a language L is a structure M= (U,I), where:

- U is a non-empty set, called the Domain, or Carrier, or Universe of Discourse of M, and

- I is an Interpretation that associates to each n-ary function symbol F of L a map

  I(F): UxU..xU -> U

  and to each n-ary predicate symbol P of L a subset of UxU..xU.

The set of functions (predicates) so introduced form the Functional Basis (Relational Basis) of the conceptualization.

Given a language L and a conceptualization (U,I), an Assignment is a map from the variables of L to U. An X-Variant of an assignment s is an assignment that is identical to s everywhere except at x where it differs.

Given a conceptualization M=(U,I) and an assignment s it is easy to extend s to map each term t of L to an individual s(t) in U by using induction on the structure of the term.

Then

- M satisfies a formula A under s iff

  o A is atomic, say P(t1 .. tn), and (s(t1) ..s(tn)) is in I(P).

  o A is (NOT B) and M does not satisfy B under s.

  o A is (B OR C) and M satisfies B under s, or M satisfies C under s. [Similarly for all other connectives.]

  o A is (ALL x B) and M satisfies B under all x-variants of s.

  o A is (EXISTS x B) and M satisfies B under some x-variants of s.

- Formula A is satisfiable in M iff there is an assignment s such that M satisfies A under s.

- Formula A is satisfiable iff there is an L-structure M such that A is satisfiable in M.

- Formula A is valid or logically true in M iff M satisfies A under any s. We then say that M is a model of A.

- Formula A is Valid or Logically True iff for any L-structure M and any assignment s, M satisfies A under s.

Some of these definitions can be made relative to a set of formulae GAMMA:

- Formula A is a Logical Consequence of GAMMA in M iff M satisfies A under any s that also satisfies all the formulae in GAMMA.

- Formula A is a Logical Consequence of GAMMA iff for any L-structure M, A is a logical consequence of GAMMA in M. At times instead of "A is a logical consequence of GAMMA" we say "GAMMA entails A".

We say that formulae A and B are (logically) equivalent iff A is a logical consequence of {B} and B is a logical consequence of {A}.

A Block World

Here we look at a problem and see how to represent it in a language. We consider a simple world of blocks as described by the following figures:

```
                                                    +--+
                                                    |a |
                                                    +--+
                                                    |e |
        +--+                                        +--+
        |a |                                        |c |
        +--+      +--+                               +--+
        |b |      |d |      ======>                 |d |
        +--+      +--+                               +--+
        |c |      |e |                               |b |
        ---------------                     --------------------
```

We see two possible states of the world. On the left is the current state, on the right a desired new state. A robot is available to do the transformation. To describe these worlds we can use a structure with domain U = {d e}, and with predicates {ON, ABOVE, CLEAR, TABLE} with the following meaning:

- ON: (ON x y) iff x is immediately above y. The interpretation of ON in the left world is {(a b) (b c) (d e)}, and in the right world is {(a e) (e c) (c d) (d b)}.

- ABOVE: (ABOVE x y) iff x is above y. The interpretation of ABOVE [in the left world] is {(a b) (b c) (a c) (d e)} and in the right world is {(a e) (a c) (a d) (a b) (e c) (e d) (e b) (c d) (c b) (d b)}

- CLEAR: (CLEAR x) iff x does not have anything above it. The interpretation of CLEAR [in the left world] is {a d} and in the right world is {a}

- TABLE: (TABLE x) iff x rests directly on the table. The interpretation of TABLE [in the left world] is {c e} and in the right world id {b}.

Examples of formulae true in the block world [both in the left and in the right state] are [these formulae are known as Non-Logical Axioms]:

- (ON x y) IMPLIES (ABOVE x y)

- ((ON x y) AND (ABOVE y z)) IMPLIES (ABOVE x z)

- (ABOVE x y) IMPLIES (NOT (ABOVE y x))

- (CLEAR x) IFF (NOT (EXISTS y (ON y x)))

- (TABLE x) IFF (NOT (EXISTS y (ON x y)))

Note that there are things that we cannot say about the block world with the current functional and predicate basis unless we use equality. Namely, we cannot say as we would like that a block can be ON at most one other block. For that we need to say that if x is ON y and x is ON z then y is z. That is, we need to use a logic with equality.

Not all formulae that are true on the left world are true on the right world and viceversa. For example, a formula true in the left world but not in the right world is (ON a b). Assertions about the left and right world can be in contradiction. For example (ABOVE b c) is true on left, (ABOVE c b) is true on right and together they contradict the non-logical axioms. This means that the theory that we have developed for talking about the block world can talk of only one world at a time. To talk about two worlds simultaneously we would need something like the Situation Calculus that we will study later.

Herbrand Structure

In first-order logic, a Herbrand structure S is a structure over a vocabulary σ, that is defined solely by the syntactical properties of σ. The idea is to take the symbols of terms as their values, e.g. the denotation of a constant symbol c is just 'c' (the symbol).

Herbrand structures play an important role in the foundations of logic programming.

Herbrand Universe

Definition

Herbrand universe will serve as the universe in *Herbrand structure.*

(1) The *Herbrand universe of a first-order language* L^σ, is the set of all ground terms of L^σ. If the language has no constants, then the language is extended by adding an arbitrary new constant.

- It is countably infinite if σ is countable and a function symbol of arity greater than 0 exists.

- In the context of first-order languages we also speak simply of the *Herbrand universe of the vocabulary σ*.

(2) The *Herbrand universe of a closed formula in Skolem normal form* F, is the set of all terms without variables, that can be constructed using the function symbols and constants of F. If F has no constants, then F is extended by adding an arbitrary new constant.

- This second definition is important in the context of computational resolution.

Example

Let L^σ be a first-order language with the vocabulary

- constant symbols: c

- function symbols: f(.), g(.) then the Herbrand universe of L^σ (or σ) is {c, f(c), g(c), f(f(c)), f(g(c)), g(f(c)), g(g(c)), ...}.

Notice that the relation symbols are not relevant for a Herbrand universe.

Herbrand Structure

A *Herbrand structure* interprets terms on top of a *Herbrand universe*.

Definition

Let S be a structure, with vocabulary σ and universe U. Let T be the set of all terms over σ and T_o be the subset of all variable-free terms. S is said to be a *Herbrand structure* iff

1. $U = T_o$

2. $f^S(t_1, ..., t_n) = f(t_1, ..., t_n)$ for every n-ary function symbol $f \in \sigma$ and $t_1, ..., t_n \in T_o$

3. $c^S = c$ for every constant $c \in \sigma$

Remarks

1. U is the Herbrand universe of σ.

2. A Herbrand structure that is a model of a theory T, is called the *Herbrand model* of T.

Examples

For a constant symbol c and a 1-ary function symbol f(.) we have the following interpretation:

- $U = \{$ c, fc, ffc, fffc, ...$\}$

- fc -> fc, ffc -> ffc, ...

- c -> c

Herbrand base

In addition to the universe, defined in *Herbrand universe*, and the term denotations, defined in *Herbrand structure*, the *Herbrand base* completes the interpretation by denoting the relation symbols.

Definition

A *Herbrand base* is the set of all ground atoms of whose argument terms are the Herbrand universe.

Examples

For a 2-ary relation symbol R, we get with the terms from above:

{ R(c, c), R(fc, c), R(c, fc), R(fc, fc), R(ffc, c), ...}

Herbrand Universe

It is a good exercise to determine for given formulae if they are satisfied/valid on specific L-structures, and to determine, if they exist, models for them. A good starting point in this task, and useful for a number of other reasons, is the Herbrand Universe for this set of formulae. Say that {Fo1 .. Fon} are the individual constants in the formulae [if there are no such constants, then introduce one, say, Fo]. Say that {F1 .. Fm} are all the non 0-ary function symbols occurring in the formulae. Then the set of (constant) terms obtained starting from the individual constants using the non 0-ary functions, is called the Herbrand Universe for these formulae.

For example, given the formula (P x A) OR (Q y), its Herbrand Universe is just {A}. Given the formulae (P x (F y)) OR (Q A), its Herbrand Universe is {A (F A) (F (F A)) (F (F (F A))) ...}.

Reduction to Clausal Form

In the following we give an algorithm for deriving from a formula an equivalent clausal form through a series of truth preserving transformations.

We can state an (unproven by us) theorem:

> *Theorem: Every formula is equivalent to a clausal form*

We can thus, when we want, restrict our attention only to such forms.

Deduction

An Inference Rule is a rule for obtaining a new formula [the consequence] from a set of given formulae [the premises].

A most famous inference rule is Modus Ponens:

{A, NOT A OR B}

B

For example:

{Sam is tall,

if Sam is tall then Sam is unhappy}

Sam is unhappy

When we introduce inference rules we want them to be Sound, that is, we want the consequence of the rule to be a logical consequence of the premises of the rule. Modus Ponens is sound. But the following rule, called Abduction , is not:

{B, NOT A OR B}

A

is not. For example:

John is wet

If it is raining then John is wet

It is raining

gives us a conclusion that is usually, but not always true [John takes a shower even when it is not raining].

A Logic or Deductive System is a language, plus a set of inference rules, plus a set of logical axioms [formulae that are valid].

A Deduction or Proof or Derivation in a deductive system D, given a set of formulae GAMMA, is a a sequence of formulae B1 B2 .. Bn such that:

- for all i from 1 to n, Bi is either a logical axiom of D, or an element of GAMMA, or is obtained from a subset of {B1 B2 .. Bi-1} by using an inference rule of D.

In this case we say that Bn is Derived from GAMMA in D, and in the case that GAMMA is empty, we say that Bn is a Theorem of D.

Soundness, Completeness, Consistency, Satisfiability

A Logic D is Sound iff for all sets of formulae GAMMA and any formula A:

- if A is derived from GAMMA in D, then A is a logical consequence of GAMMA

A Logic D is Complete iff for all sets of formulae GAMMA and any formula A:

- If A is a logical consequence of GAMMA, then A can be derived from GAMMA in D.

A Logic D is Refutation Complete iff for all sets of formulae GAMMA and any formula A:

- If A is a logical consequence of GAMMA, then the union of GAMMA and (NON A) is inconsistent

Note that if a Logic is Refutation Complete then we can enumerate all the logical consequences of GAMMA and, for any formula A, we can reduce the question if A is or not a logical consequence of GAMMA to the question: the union of GAMMA and NOT A is or not consistent.

We will work with logics that are both Sound and Complete, or at least Sound and Refutation Complete.

A Theory T consists of a logic and of a set of Non-logical axioms. For convenience, we may refer, when not ambiguous, to the logic of T, or the non-logical axioms of T, just as T.

The common situation is that we have in mind a well defined "world" or set of worlds. For example we may know about the natural numbers and the arithmetic operations and relations. Or we may think of the block world. We choose a language to talk about these worlds. We introduce function and predicate symbols as it is appropriate. We then introduce formulae, called Non-Logical Axioms, to characterize the things that are true in the worlds of interest to us. We choose a logic, hopefully sound and (refutation) complete, to derive new facts about the worlds from the non-logical axioms.

A Theorem in a theory T is a formula A that can be derived in the logic of T from the non-logical axioms of T.

A Theory T is Consistent iff there is no formula A such that both A and NOT A are theorems of T; it is Inconsistent otherwise. If a theory T is inconsistent, then, for essentially any logic, any formula is a theorem of T. [Since T is inconsistent, there is a formula A such that both A and NOT A are theorems of T. It is hard to imagine a logic where from A and (NOT A) we cannot infer FALSE, and from FALSE we cannot infer any formula. We will say that a logic that is at least this powerful is Adeguate.]

A Theory T is Unsatisfiable if there is no structure where all the non-logical axioms of T are valid. Otherwise it is Satisfiable.

Given a Theory T, a formula A is a Logical Consequence of T if it is a logical consequence of the non logical axioms of T.

Theorem: If the logic we are using is sound then:

1. If a theory T is satisfiable then T is consistent

2. If the logic used is also adequate then if T is consistent then T is satisfiable

3. If a theory T is satisfiable and by adding to T the non-logical axiom (NOT A) we get a theory that is not satisfiable Then A is a logical consequence of T.

4. If a theory T is satisfiable and by adding the formula (NOT A) to T we get a theory that is inconsistent, then A is a logical consequence of T.

Resolution (Logic)

In mathematical logic and automated theorem proving, resolution is a rule of inference leading to a refutation theorem-proving technique for sentences in propositional logic and first-order logic. In other words, iteratively applying the resolution rule in a suitable way allows for telling whether a propositional formula is satisfiable and for proving that a first-order formula is unsatisfiable. Attempting to prove a satisfiable first-order formula as unsatisfiable may result in a nonterminating computation; this problem doesn't occur in propositional logic.

The resolution rule can be traced back to Davis and Putnam (1960); however, their algorithm required to try all ground instances of the given formula. This source of combinatorial explosion was eliminated in 1965 by John Alan Robinson's syntactical unification algorithm, which allowed one to instantiate the formula during the proof "on demand" just as far as needed to keep refutation completeness.

The clause produced by a resolution rule is sometimes called a resolvent.

Resolution in Propositional Logic

Resolution Rule

The resolution rule in propositional logic is a single valid inference rule that produces a new clause implied by two clauses containing complementary literals. A literal is a propositional variable or the negation of a propositional variable. Two literals are said to be complements if one is the negation of the other (in the following, $\neg c$ is taken to be the complement to c). The resulting clause contains all the literals that do not have complements. Formally:

$$\frac{a_1 \vee \ldots \vee a_{i-1} \vee c \vee a_{i+1} \vee \ldots \vee a_n, \quad b_1 \vee \ldots \vee b_{j-1} \vee \neg c \vee b_{j+1} \vee \ldots \vee b_m}{a_1 \vee \ldots \vee a_{i-1} \vee a_{i+1} \vee \ldots \vee a_n \vee b_1 \vee \ldots \vee b_{j-1} \vee b_{j+1} \vee \ldots \vee b_m}$$

where

> all as , bs and cs are literals,
>
> the dividing line stands for entails

The clause produced by the resolution rule is called the *resolvent* of the two input clauses. It is the principle of *consensus* applied to clauses rather than terms.

When the two clauses contain more than one pair of complementary literals, the resolution rule can be applied (independently) for each such pair; however, the result is always a tautology.

Modus ponens can be seen as a special case of resolution (of a one-literal clause and a two-literal clause).

$$\frac{p \rightarrow q, p}{q}$$

is equivalent to

$$\frac{\neg p \vee q, p}{q}$$

A Resolution Technique

When coupled with a complete search algorithm, the resolution rule yields a sound and complete algorithm for deciding the *satisfiability* of a propositional formula, and, by extension, the validity of a sentence under a set of axioms.

This resolution technique uses proof by contradiction and is based on the fact that any sentence in propositional logic can be transformed into an equivalent sentence in conjunctive normal form. The steps are as follows.

- All sentences in the knowledge base and the *negation* of the sentence to be proved (the *conjecture*) are conjunctively connected.

- The resulting sentence is transformed into a conjunctive normal form with the conjuncts viewed as elements in a set, S, of clauses.

  o For example $(A_1 \vee A_2) \wedge (B_1 \vee B_2 \vee B_3) \wedge (C_1)$ gives rise to the set $S = \{A_1 \vee A_2, B_1 \vee B_2 \vee B_3, C_1\}$.

- The resolution rule is applied to all possible pairs of clauses that contain complementary literals. After each application of the resolution rule, the resulting sentence is simplified by removing repeated literals. If the sentence contains complementary literals, it is discarded (as a tautology). If not, and if it is not yet present in the clause set S, it is added to S, and is considered for further resolution inferences.

- If after applying a resolution rule the *empty clause* is derived, the original formula is unsatisfiable (or *contradictory*), and hence it can be concluded that the initial conjecture follows from the axioms.

- If, on the other hand, the empty clause cannot be derived, and the resolution rule cannot be applied to derive any more new clauses, the conjecture is not a theorem of the original knowledge base.

One instance of this algorithm is the original Davis–Putnam algorithm that was later refined into the DPLL algorithm that removed the need for explicit representation of the resolvents.

This description of the resolution technique uses a set S as the underlying data-structure to represent resolution derivations. Lists, Trees and Directed Acyclic Graphs are other possible and common alternatives. Tree representations are more faithful to the fact that the resolution rule is binary. Together with a sequent notation for clauses, a tree representation also makes it clear to see how the resolution rule is related to a special case of the cut-rule, restricted to atomic cut-formulas. However, tree representations are not as compact as set or list representations, because they explicitly show redundant subderivations of clauses that are used more than once in the derivation of the empty clause. Graph representations can be as compact in the number of clauses as list representations and they also store structural information regarding which clauses were resolved to derive each resolvent.

A Simple Example

$$\frac{a \vee b, \quad \neg a \vee c}{b \vee c}$$

In plain language: Suppose a is false. In order for the premise $a \vee b$ to be true, b must be true. Alternatively, suppose a is true. In order for the premise $\neg a \vee c$ to be true, c must be true. Therefore regardless of falsehood or veracity of a, if both premises hold, then the conclusion $b \vee c$ is true.

Resolution in First Order Logic

Resolution rule can be generalized to first-order logic to:

$$\frac{\Gamma_1 \cup \{L_1\} \; \Gamma_2 \cup \{L_2\}}{(\Gamma_1 \cup \Gamma_2)\phi}\phi$$

where ϕ is a most general unifier of L_1 and $\overline{L_2}$ and Γ_1 and Γ_2 have no common variables.

Example

The clauses $P(x), Q(x)$ and $\neg P(b)$ can apply this rule with $[b/x]$ as unifier.

Here x is a variable and b is a constant.

$$\frac{P(x), Q(x) \; \neg P(b)}{Q(b)}[b/x]$$

Here we see that

- The clauses $P(x), Q(x)$ and $\neg P(b)$ are the inference's premises
- $Q(b)$ (the resolvent of the premises) is its conclusion.
- The literal $P(x)$ is the left resolved literal,
- The literal $\neg P(b)$ is the right resolved literal,
- P is the resolved atom or pivot.
- $[b/x]$ is the most general unifier of the resolved literals.

Informal Explanation

In first order logic, resolution condenses the traditional syllogisms of logical inference down to a single rule.

To understand how resolution works, consider the following example syllogism of term logic:

> All Greeks are Europeans.
>
> Homer is a Greek.
>
> Therefore, Homer is a European.

Or, more generally:

> $\forall x.P(x) \Rightarrow Q(x)$
>
> $P(a)$
>
> Therefore, $Q(a)$

To recast the reasoning using the resolution technique, first the clauses must be converted to conjunctive normal form (CNF). In this form, all quantification becomes implicit: universal quantifiers on variables (X, Y, ...) are simply omitted as understood, while existentially-quantified variables are replaced by Skolem functions.

$$\neg P(x) \lor Q(x)$$

$$P(a)$$

Therefore, $Q(a)$

So the question is, how does the resolution technique derive the last clause from the first two? The rule is simple:

- Find two clauses containing the same predicate, where it is negated in one clause but not in the other.

- Perform a unification on the two predicates. (If the unification fails, you made a bad choice of predicates. Go back to the previous step and try again.)

- If any unbound variables which were bound in the unified predicates also occur in other predicates in the two clauses, replace them with their bound values (terms) there as well.

- Discard the unified predicates, and combine the remaining ones from the two clauses into a new clause, also joined by the "$\lor$" operator.

To apply this rule to the above example, we find the predicate P occurs in negated form

$$\neg P(X)$$

in the first clause, and in non-negated form

$$P(a)$$

in the second clause. X is an unbound variable, while a is a bound value (term). Unifying the two produces the substitution

$$X \mapsto a$$

Discarding the unified predicates, and applying this substitution to the remaining predicates (just $Q(X)$, in this case), produces the conclusion:

$$Q(a)$$

For another example, consider the syllogistic form

All Cretans are islanders.

All islanders are liars.

Therefore all Cretans are liars.

Or more generally,

$$\forall X\, P(X) \rightarrow Q(X)$$

$$\forall X\, Q(X) \rightarrow R(X)$$

Therefore, $\forall X\, P(X) \rightarrow R(X)$

In CNF, the antecedents become:

$$\neg P(X) \lor Q(X)$$

$$\neg Q(Y) \lor R(Y)$$

(Note that the variable in the second clause was renamed to make it clear that variables in different clauses are distinct.)

Now, unifying $Q(X)$ in the first clause with $\neg Q(Y)$ in the second clause means that X and Y become the same variable anyway. Substituting this into the remaining clauses and combining them gives the conclusion:

$$\neg P(X) \lor R(X)$$

The resolution rule, as defined by Robinson, also incorporated factoring, which unifies two literals in the same clause, before or during the application of resolution as defined above. The resulting inference rule is refutation-complete, in that a set of clauses is unsatisfiable if and only if there exists a derivation of the empty clause using resolution alone.

Paramodulation

Paramodulation is a related technique for reasoning on sets of clauses where the predicate symbol is equality. It generates all "equal" versions of clauses, except reflexive identities. The paramodulation operation takes a positive *from* clause, which must contain an equality literal. It then searches an *into* clause with a subterm that unifies with one side of the equality. The subterm is then replaced by the other side of the equality. The general aim of paramodulation is to reduce the system to atoms, reducing the size of the terms when substituting.

Implementations

- CARINE
- Gandalf
- Otter
- Prover9
- SNARK
- SPASS
- Vampire

Resolution

We have introduced the inference rule Modus Ponens. Now we introduce another inference rule that is particularly significant, Resolution.

Since it is not trivial to understand, we proceed in two steps. First we introduce Resolution in the Propositional Calculus, that is, in a language with only truth valued variables. Then we generalize to First Order Logic.

Resolution in the Propositional Calculus

In its simplest form Resolution is the inference rule:

$$\frac{\{A\ OR\ C,\ B\ OR\ (NOT\ C)\}}{A\ OR\ B}$$

More in general the Resolution Inference Rule is:

- Given as premises the clauses C1 and C2, where C1 contains the literal L and C2 contains the literal (NOT L), infer the clause C, called the Resolvent of C1 and C2, where C is the union of (C1 - {L}) and (C2 -{(NOT L)})

In symbols:

$$\frac{\{C1,\ C2\}}{(C1 - \{L\})\ UNION\ (C2 - \{(NOT\ L)\})}$$

Example:

The following set of clauses is inconsistent:

1. (P OR (NOT Q))

2. ((NOT P) OR (NOT S))

3. (S OR (NOT Q))

4. Q

In fact:

5. ((NOT Q) OR (NOT S))   from 1. and 2.

6. (NOT Q)   from 3. and 5.

7. FALSE   from 4. and 6.

Notice that 7. is really the empty clause [why?].

Theorem: The Propositional Calculus with the Resolution Inference Rule is sound and Refutation Complete.

NOTE: This theorem requires that clauses be represented as sets, that is, that each element of the clause appear exactly once in the clause. This requires some form of membership test when elements are added to a clause.

$C_1 = \{P\ P\}$

$\quad C_2 = \{(NOT\ P)\ (NOT\ P)\}$

$\quad C = \{P\ (NOT\ P)\}$

From now on by resolution we just get again C_1, or C_2, or C.

Resolution in First Order Logic

Given clauses C_1 and C_2, a clause C is a RESOLVENT of C_1 and C_2, if

1. There is a subset $C_1' = \{A_1, .., A_m\}$ of C_1 of literals of the same sign, say positive, and a subset $C_2' = \{B_1, .., B_n\}$ of C_2 of literals of the opposite sign, say negative,

2. There are substitutions s_1 and s_2 that replace variables in C_1' and C_2' so as to have new variables,

3. C_2'' is obtained from C_2 removing the negative signs from $B_1 .. B_n$

4. There is an Most General Unifier s for the union of $C_1'.s_1$ and $C_2''.s_2$

and C is

$\quad ((C_1 - C_1').s_1\ UNION\ (C_2 - C_2').s_2).s$

In symbols this Resolution inference rule becomes:

$$\frac{\{C_1,\ C_2\}}{C}$$

If C_1' and C_2' are singletons (i.e. contain just one literal), the rule is called Binary Resolution.

Example:

$\quad C_1 = \{(P\ z\ (F\ z))\ (P\ z\ A)\}$

$\quad C_2 = \{(NOT\ (P\ z\ A))\ (NOT\ (P\ z\ x))\ (NOT\ (P\ x\ z))\}$

$\quad C_1' = \{(P\ z\ A)\}$

$\quad C_2' = \{(NOT\ (P\ z\ A))\ (NOT\ (P\ z\ x))\}$

$\quad C_2'' = \{(P\ z\ A)\ (P\ z\ x)\}$

$s1 = [z1/z]$

$s2 = [z2/z]$

C1'.s1 UNION C2'.s2 = {(P z1 A) (P z2 A) (P z2 x)}

$s = [z1/z2 \ A/x]$

C = {(NOT (P A z1)) (P z1 (F z1))}

Notice that this application of Resolution has eliminated more than one literal from C2, i.e. it is not a binary resolution.

Theorem: First Order Logic, with the Resolution Inference Rule, is sound and refutation complete.

We will not develop the proof of this theorem. We will instead look at some of its steps, which will give us a wonderful opportunity to revisit Herbrand. But before that let's observe that in a sense, if we replace in this theorem "Resolution" by "Binary Resolution", then the theorem does not hold and Binary Resolution is not Refutation Complete. This is the case when in the implementation we do not use sets but instead use bags. This can be shown using the same example as in the case of propositional logic.

Given a clause C, a subset D of C, and a substitution s that unifies D, we define C.s to be a Factor of C. The Factoring Inference Rule is the rule with premise C and as consequence C.s.

> *Theorem: For any set of clauses S and clause C, if C is derivable from S using Resolution, then C is derivable from S using Binary Resolution and Factoring.*

When doing proofs it is efficient to have as few clauses as possible. The following definition and rule are helpful in eliminating redundant clauses:

> A clause C1 Subsumes a clause C2 iff there is a substitution s such that C1.s is a subset of C2.

> Subsumption Elimination Rule: If C1 subsumes C2 then eliminate C2.

Herbrand Revisited

We have presented the concept of Herbrand Universe H_s for a set of clauses S. Here we meet the concept of Herbrand Base, H(S), for a set of clauses S. H(S) is obtained from S by considering the ground instances of the clauses of S under all the substitutions that map all the variables of S into elements of the Herbrand universe of S. Clearly, if in S occurs some variable and the Herbrand universe of S is infinite then H(S) is infinite. [NOTE: Viceversa, if S has no variables, or S has variables and possibly individual constants, but no other function symbol, then H(S) is finite. If H(S) is finite then we can, as we will see, decide if S is or not satisfiable.] [NOTE: it is easy to determine if a finite subset of H(S) is satisfiable: since it consists of ground clauses, the truth table method works now as in propositional cases.]

The importance of the concepts of Herbrand Universe and of Herbrand Base is due to the following theorems:

Herbrand Theorem: If a set S of clauses is unsatisfiable then there is a finite subset of H(S) that is also unsatisfiable.

Because of the theorem, when H(S) is finite we will be able to decide is S is or not satisfiable. Herbrand theorem immediately suggests a general refutation complete proof procedure:

given a set of clauses S, enumerate subsets of H(S) until you find one that is unsatisfiable.

But, as we shall soon see, we can come up with a better refutation complete proof procedure.

Refutation Completeness of the Resolution Proof Procedure

Given a set of clauses S, the Resolution Closure of S, R(S), is the smallest set of clauses that contains S and is closed under Resolution. In other words, it is the set of clauses obtained from S by applying repeatedly resolution.

Ground Resolution Theorem: If S is an unsatisfiable set of ground clauses, then R(S) contains the clause FALSE.

In other words, there is a resolution deduction of FALSE from S.

Lifting Lemma: Given clauses C1 and C2 that have no variables in common, and ground instances C1' and C2', respectively, of C1 and C2, if C' is a resolvent of C1' and C2', then there is a clause C which is a resolvent of C1 and C2 which has C' as a ground instance

With this we have our result, that the Resolution Proof procedure is Refutation Complete. Note the crucial role played by the Herbrand Universe and Basis. Unsatisfiability of S is reduced to unsatisfiability for a finite subset $H_s(S)$ of H(S), which in turn is reduced to the problem of finding a resolution derivation for FALSE in $H_s(S)$, derivation which can be "lifted" to a resolution proof of FALSE from S.

Dealing with Equality

Up to now we have not dealt with equality, that is, the ability to recognize terms as being equivalent (i.e. always denoting the same individual) on the basis of some equational information. For example, given the information that

S(x) = x+1

then we can unify:

F(S(x) y) and F(x+1, 3).

There are two basic approaches to dealing with this problem.

- The first is to add inference rules to help us replace terms by equal terms. One such rule is the Demodulation Rule: Given terms t1, t2, and t3 where t1 and t2 are unifiable with MGU s, and t2 occurs in a formula A, then

$$\frac{\{t_1 = t_3, A(\ldots\, t_2\, \ldots)\}}{A(\ldots\, t_3.s\, \ldots)}$$

Another more complex, and useful, rule is Paramodulation.

- The second approach is not to add inference rules and instead to add non-logical axioms that characterize equality and its effect for each non logical symbol. We first establish the reflexive, symmetric and transitive properties of "=": x=x

  x=y IMPLIES y=x

  x=y AND y=z IMPLIES x=z

  Then for each unary function symbol F we add the equality axiom

  x=y IMPLIES F(x)=F(y)

  Then for each binary function symbol F we add the equality axiom

  x=z AND y=w IMPLIES F(x y)=F(z w)

  And similarly for all other function symbols.

  The treatment of predicate symbols is similar. For example, for the binary predicate symbol P we add

  x=z AND y=w IMPLIES (P(x y) IFF P(z w))

Whatever approach is chosen, equality substantially complicates proofs.

Answering True/False Questions

If we want to show that a clause C is derivable from a set of clauses S={C1 C2 .. Cn}, we add to S the clauses obtained by negating C, and apply resolution to the resulting set S' of clauses until we get the clause FALSE.

Example

We are back in the Block World with the following state

```
+--+
|C |
+--+      +--+
|A |      |B |
----+--+----+--+-------
```

which gives us the following State Clauses:

- ON(C A)

- ONTABLE(A)

- ONTABLE(B)

- CLEAR(C)

- CLEAR(B)

In addition we consider the non-logical axiom:
- (ALL x (CLEAR(x) IMPLIES (NOT (EXISTS y ON(y x)))))

which in clause form becomes

- NOT CLEAR(x) OR NOT ON(y x)

If we now ask whether (NOT (EXISTS y (ON(y C)))), we add to the clauses considered above the clause ON(F C) and apply resolution:

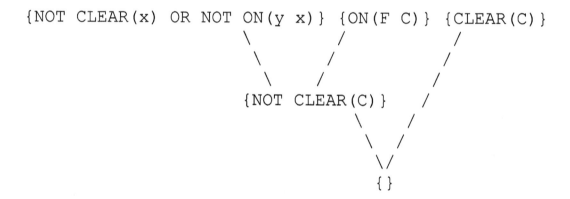

Example

We are given the following predicates:

- S(x) : x is Satisfied

- H(x) : x is Healthy

- R(x) : x is Rich

- P(x) : x is Philosophical

The premises are the non-logical axioms:

- S(x) IMPLIES (H(x) AND R(x))

- EXISTS x (S(x) and P(x))

The conclusion is

- EXISTS x (P(x) AND R(x))

The corresponding clauses are:

1. NOT S(x) OR H(x)

2. NOT S(x) OR R(x)

3. S(B)

4. P(B)

5. NOT P(x) OR NOT R(x)

where B is a Skolem constant.

The proof is then:

The proof is then:

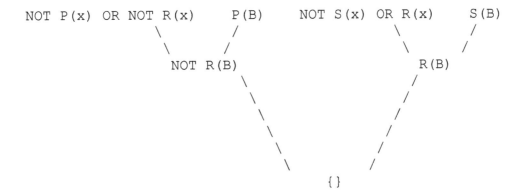

Answering Fill-in-Blanks Questions

We now determine how we can identify individual(s) that satisfy specific formulae.

EXAMPLE:

NON-LOGICAL SYMBOLS:

- SW(x y): x is staying with y

- A(x y): x is at place y

- R(x y): x can be reached at phone number y

- PH(x): the phone number for place x

- Sally, Morton, UnionBldg: Individuals

NON-LOGICAL AXIOMS:

1. SW(Sally Morton)

2. A(Morton UnionBlidg)

3. SW(x y) AND A(y z) IMPLIES A(x z), which is equivalent to the clause

  1. NOT SW(x y) OR NOT A(y z) OR A(x z)

4. A(x y) IMPLIES R(x PH(y)), which is equivalent to the clause

  1. NOT A(u v) OR R(u PH(v))

GOAL: Determine where to call Sally

- NOT EXISTS x R(Sally x), which is equivalent to the clause

  1. NOT R(Sally w).

To this clause we add as a disjunct the literal, Answer Literal, Ans(w) to obtain the clause :

5. Ans(w) OR NOT R(Sally w).

PROOF

6. Ans(v) OR NOT A(Sally v). from 4. and 5.

7. Ans(v) OR NOT SW(Sally y) OR NOT A(y v), from 6. and 3.

8. Ans(v) OR NOT A(Morton v), from 7. and 1.

9. Ans(UnionBldg), from 8. and 2.

The proof procedure terminates when we get a clause that is an instance of the Answer Literal. 9. and gives us the place where we can call Sally.

General Method

If A is the Fill-In-Blanks question that we need to answer and x1 .. xn are the free variables occurring in A, then we add to the Non-Logical axioms and Facts GAMMA the clause

  NOT A OR ANS(x1 .. xn)

We terminate the proof when we get a clause of the form

  ANS(t1 .. tn)

t1 .. tn are terms that denote individuals that simultaneously satisfy the query for, respectively x1 .. xn.

We can obtain all the individuals that satisfy the original query by continuing the proof looking for alternative instantiations for the variables x1 .. xn.

If we build the proof tree for ANS(t1 .. tn) and consider the MGUs used in it, the composition of these substitutions, restricted to x1 .. xn, gives us the individuals that answer the original Fill-In-Blanks question.

Rule Based Systems

Instead of representing knowledge in a relatively declarative, static way (as a bunch of things that are true), rule-based system represent knowledge in terms of a bunch of rules that tell you what you should do or what you could conclude in different situations. A rule-based system consists of a bunch of IF-THEN rules, a bunch of facts, and some interpreter controlling the application of the rules, given the facts. Hence, this are also sometimes referred to as production systems. Such rules can be represented using Horn clause logic.

There are two broad kinds of rule system: forward chaining systems, and backward chaining systems. In a forward chaining system you start with the initial facts, and keep using the rules to draw new conclusions (or take certain actions) given those facts. In a backward chaining system you start with some hypothesis (or goal) you are trying to prove, and keep looking for rules that would allow you to conclude that hypothesis, perhaps setting new subgoals to prove as you go. Forward chaining systems are primarily data-driven, while backward chaining systems are goal-driven. We'll look at both, and when each might be useful.

Horn Clause Logic

There is an important special case where inference can be made substantially more focused than in the case of general resolution. This is the case where all the clauses are *Horn clauses*.

Definition: A *Horn clause* is a clause with at most one positive literal.

Any Horn clause therefore belongs to one of four categories:

- A *rule*: 1 positive literal, at least 1 negative literal. A rule has the form "$\sim P_1 \vee \sim P_2 \vee ... \vee \sim P_k \vee Q$". This is logically equivalent to "$[P_1 \wedge P_2 ... \wedge P_k] \Rightarrow Q$"; thus, an if-then implication with any number of conditions but one conclusion. Examples: "$\sim man(X) \vee mortal(X)$" (All men are mortal); "$\sim parent(X,Y) \vee \sim ancestor(Y,Z) \vee ancestor(X,Z)$" (If X is a parent of Y and Y is an ancestor of Z then X is an ancestor of Z.)

- A *fact* or *unit*: 1 positive literal, 0 negative literals. Examples: "$man(socrates)$", "$parent(elizabeth,charles)$", "$ancestor(X,X)$" (Everyone is an ancestor of themselves (in the trivial sense).)

- A *negated goal* : 0 positive literals, at least 1 negative literal. In virtually all implementations of Horn clause logic, the negated goal is the negation of the statement to be proved; the knowledge base consists entirely of facts and goals. The statement to be proven, therefore, called the goal, is therefore a single unit or the conjuction of units; an *existentially* quantified variable in the goal turns into a *free* variable in the negated goal. E.g. If the goal to be proven is "$exists (X) male(X)ancestor(elizabeth,X)$" (show that there exists a male descendent of Elizabeth) the negated goal will be "$\sim male(X) \vee \sim ancestor(elizabeth,X)$".

- The null clause: 0 positive and 0 negative literals. Appears only as the end of a resolution proof.

Now, if resolution is restricted to Horn clauses, some interesting properties appear. Some of these are evident; others I will just state and you can take on faith.

I. If you resolve Horn clauses A and B to get clause C, then the positive literal of A will resolve against a negative literal in B, so the only positive literal left in C is the one from B (if any). Thus, the resolvent of two Horn clauses is a Horn clause.

II. If you resolve a negated goal G against a fact or rule A to get clause C, the positive literal in A resolves against a negative literal in G. Thus C has no positive literal, and thus is either a negated goal or the null clause.

III. Therefore: Suppose you are trying to prove Phi from Gamma, where ~Phi is a negated goal, and Gamma is a knowledge base of facts and rules. Suppose you use the set of support strategy, in which no resolution ever involves resolving two clauses from Gamma together. Then, inductively, every resolution combines a negated goal with a fact or rule from Gamma and generates a new negated goal. Moreover, if you take a resolution proof, and trace your way back from the null clause at the end to ~Phi at the beginning, since every resolution involves combining one negated goal with one clause from Gamma, it is clear that the sequence of negated goals involved can be linearly ordered. That is, the final proof, ignoring dead ends has the form

```
~Phi resolves with C1 from Gamma, generating negated goal P2

P2 resolves with C2 from Gamma, generating negated goal P3

...
```

Pk resolves with C2 from Gamma, generating the null clause.

IV. Therefore, the process of generating the null clause can be viewed as a state space search where:

- A state is a negated goal.

- A operator on negated goal P is to resolve it with a clause C from Gamma.

- The start state is ~Phi

- The goal state is the null clause.

V. Moreover, it turns out that it doesn't really matter which literal in P you choose to resolve. All the literals in P will have to be resolved away eventually, and the order doesn't really matter. (This takes a little work to prove or even to state precisely, but if you work through a few examples, it becomes reasonably evident.)

Backward Chaining

Putting all the above together, we formulate the following non-deterministic algorithm for resolution in Horn theories. This is known as backward chaining.

```
bc(in P0 : negated goal;

    GAMMA : set of facts and rules;)
```

```
{ if P0 = null then succeed;

  pick a literal L in P0;

  choose a clause C in GAMMA whose head resolves with L;

  P := resolve(P0,GAMMA);

  bc(P,GAMMA)

}
```

If bc(~Phi,Gamma) succeeds, then Phi is a consequence of Gamma; if it fails, then Phi is not a consequence of Gamma.

Moreover: Suppose that Phi contains existentially quantified variables. As remarked above, when ~Phi is Skolemized, these become free variables. If you keep track of the successive bindings through the successful path of resolution, then the final bindings of these variables gives you a *value* for these variables; all proofs in Horn theories are constructive (assuming that function symbols in Gamma are constructive.) Thus the attempt to prove a statement like "exists(X,Y) p(X,Y)^q(X,Y)" can be interpreted as " *Find* X and Y such that p(X,Y) and q(X,Y)."

The succcessive negated goals Pi can be viewed as negations of *subgoals* of Phi. Thus, the operation of resolving ~P against C to get ~Q can be interpreted, "One way to prove P would be to prove Q and then use C to infer P". For instance, suppose P is "mortal(socrates)," C is "man(X) => mortal(X)" and Q is "man(socrates)." Then the step of resolving ~P against C to get ~Q can be viewed as, "One way to prove mortal(socrates) would to prove man(socrates) and then combine that with C."

Propositional Horn theories can be decided in polynomial time. First-order Horn theories are only semi-decidable, but in practice, resolution over Horn theories runs much more efficiently than resolution over general first-order theories, because of the much restricted search space used in the above algorithm.

Backward chaining is complete for Horn clauses. If Phi is a consequence of Gamma, then there is a backward-chaining proof of Phi from Gamma.

Pure Prolog

We are now ready to deal with (pure) Prolog, the major Logic Programming Language. It is obtained from a variation of the backward chaining algorithm that allows Horn clauses with the following rules and conventions:

- The Selection Rule is to select the leftmost literals in the goal.

- The Search Rule is to consider the clauses in the order they appear in the current list of clauses, from top to bottom.

- Negation as Failure, that is, Prolog assumes that a literal L is proven if if it is unable to prove (NOT L)

- Terms can be set equal to variables but not in general to other terms. For example, we can say that x=A and x=F(B) but we cannot say that A=F(B).

- Resolvents are added to the bottom of the list of available clauses.

These rules make for very rapid processing. Unfortunately:

The Pure Prolog Inference Procedure is Sound but not Complete

This can be seen by example. Given

- $P(A,B)$

- $P(C,B)$

- $P(y,x) < = P(x,y)$

- $P(x,z) < = P(x,y),P(y,z)$

we are unable to derive in Prolog that $P(A,C)$ because we get caught in an ever deepening depth-first search.

A Prolog "program" is a knowledge base Gamma. The program is invoked by posing a query Phi. The value returned is the bindings of the variables in Phi, if the query succeeds, or failure. The interpreter returns one answer at a time; the user has the option to request it to continue and to return further answers.

The derivation mechanism of Pure Prolog has a very simple form that can be described by the following flow chart.

Interpreter for Pure Prolog

Notational conventions:

- i: used to index literals in a goal

- K_i: indexes the clauses in the given program (i.e. set of clauses) P

- Max: the number of clauses in P

- h(G): the first literal of the goal G

- t(G): the rest of goal G, i.e. G without its first literal

- clause(K_i): the kith clause of the program

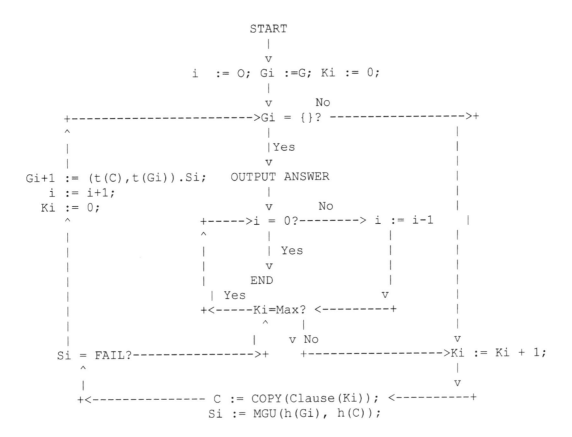

Real Prolog

Real Prolog systems differ from pure Prolog for a number of reasons. Many of which have to do with the ability in Prolog to modify the control (search) strategy so as to achieve efficient programs. In fact there is a dictum due to Kowalski:

Logic + Control = Algorithm

But the reason that is important to us now is that Prolog uses a Unification procedure which does not enforce the Occur Test. This has an unfortunate consequence that, while Prolog may give origin to efficient programs, but

Prolog is not Sound

Actual Prolog differs from pure Prolog in three major respects:

- There are additional functionalities besides theorem proving, such as functions to assert statements, functions to do arithmetic, functions to do I/O.

- The "cut" operator allows the user to prune branches of the search tree.

- The unification routine is not quite correct, in that it does not check for circular bindings e.g. X -> Y, Y -> f(X).)

Notation: The clause "~p V ~q V r" is written in Prolog in the form "r :- p,q."

Example: Let Gamma be the following knowledge base:

```
1. ancestor(X,X).

2. ancestor(X,Z) :- parent(X,Y), ancestor(Y,Z).

3. parent(george,sam).

4. parent(george,andy).

5. parent(andy,mary).

6. male(george).

7. male(sam).

8. male(andy).

9. female(mary).
```

Let Phi be the query "exists(Q) ancestor(george,Q)female(Q)." (i.e. find a female descendant of george.) Then the Skolemization of Phi is "~ancestor(george,Q) V ~female(Q)." A Prolog search proceeds as follows: (The indentation indicates the subgoal structure. Note that the variables in clauses in Gamma are renamed each time.)

```
G0: ~ancestor(george,Q) V ~female(Q). Resolve with 1: Q=X1=george.

  G1: ~female(george) Fail. Return to G0.

G0: ~ancestor(george,Q)  V  ~female(Q).  Resolve  with  2:  X2=george.
Z2=Q.

  G2: ~parent(george,Y2) V ~ancestor(Y2,Q) V ~female(Q).

                                 Resolve with 3: Y2=sam.

    G3: ~ancestor(sam,Q) V ~female(Q). Resolve with 1: Q=X3=sam.

      G4: ~female(sam).     Fail. Return to G3.

    G3: ~ancestor(sam,Q) V ~female(Q). Resolve with 2: X4=sam, Z4=Q

      G5: ~parent(sam,Y2) V ~ancestor(Y2,Q) V ~female(Q).

                                 Fail. Return to G3.

    G3: ~ancestor(sam,Q) V ~female(Q). Fail. Return to G2.

  G2: ~parent(george,Y2) V ~ancestor(Y2,Q) V ~female(Q).

                                 Resolve with 4: Y2=andy.

    G6: ~ancestor(andy,Q) V ~female(Q). Resolve with 1: X5=Q=andy

      G7: ~female(andy). Fail. Return to G6.
```

G6: ~parent(andy,Y6) V ~ancestor(Y6,Q) V ~female(Q).

Resolve with 5: Y6=mary.

G8: ~ancestor(mary,Q) V ~female(mary). Resolve with 1: X7=Q=mary.

G9: ~female(mary) Resolve with 9.

Null.

Return the binding Q=mary.

Forward Chaining

An alternative mode of inference in Horn clauses is *forward chaining* . In forward chaining, one of the resolvents in every resolution is a fact. (Forward chaining is also known as "unit resolution.")

Forward chaining is generally thought of as taking place in a dynamic knowledge base, where facts are gradually added to the knowledge base Gamma. In that case, forward chaining can be implemented in the following routines.

```
procedure add_fact(in F; in out GAMMA)

    /* adds fact F to knowledge base GAMMA and forward chains */

if F is not in GAMMA then {

    GAMMA := GAMMA union {F};

    for each rule R in GAMMA do {

        let ~L be the first negative literal in R;

        if L unifies with F then

          then { resolve R with F to get C;

                if C is a fact then add_fact(C,GAMMA)

                else /* C is a rule */ add_rule(C,GAMMA)

                }

        }

    }

end add_fact.

procedure add_rule(in R; in out GAMMA)

    /* adds rule R to knowledge base GAMMA and forward chains */
```

```
if R is not in GAMMA then {

    GAMMA := GAMMA union {R};

    let ~L be the first negative literal in R;

    for each fact F in GAMMA do

        if L unifies with F

          then { resolve R with F to get C;

                    if C is a fact then add_fact(C,GAMMA)

                    else /* C is a rule */ add_rule(C,GAMMA)

                }

    }

end add_fact.

procedure answer_query(in Q, GAMMA) return boolean /* Success or failure
*/

{ QQ := {Q} /* A queue of queries

    while QQ is non-empty do {

        Q1 := pop(QQ);

        L1 := the first literal in Q1;

        for each fact F in GAMMA do

            if F unifies with L

              then { resolve F with Q1 to get Q2;

                        if Q2=null then return(true)

                              else add Q2 to QQ;

                    }

        }

return(false)
```

The forward chaining algorithm may not terminate if GAMMA contains recursive rules.

Forward chaining is complete for Horn clauses; if Phi is a consequence of Gamma, then there is a forward chaining proof of Phi from Gamma. To be sure of finding it if Gamma contains recursive rules, you have to modify the above routines to use an exhaustive search technique, such as a breadth-first search.

In a forward chaining system the facts in the system are represented in a *working memory* which is continually updated. Rules in the system represent possible actions to take when specified conditions hold on items in the working memory - they are sometimes called condition-action rules. The conditions are usually *patterns* that must *match* items in the working memory, while the actions usually involve *adding* or *deleting* items from the working memory.

The interpreter controls the application of the rules, given the working memory, thus controlling the system's activity. It is based on a cycle of activity sometimes known as a *recognise-act* cycle. The system first checks to find all the rules whose conditions hold, given the current state of working memory. It then selects one and performs the actions in the action part of the rule. (The selection of a rule to fire is based on fixed strategies, known as *conflict resolution* strategies.) The actions will result in a new working memory, and the cycle begins again. This cycle will be repeated until either no rules fire, or some specified goal state is satisfied.

Conflict Resolution Strategies

A number of conflict resolution strategies are typically used to decide which rule to fire. These include:

- Don't fire a rule twice on the same data.

- Fire rules on more recent working memory elements before older ones. This allows the system to follow through a single chain of reasoning, rather than keeping on drawing new conclusions from old data.

- Fire rules with more specific preconditions before ones with more general preconditions. This allows us to deal with non-standard cases. If, for example, we have a rule ``IF (bird X) THEN ADD (flies X)'' and another rule ``IF (bird X) AND (penguin X) THEN ADD (swims X)'' and a penguin called tweety, then we would fire the second rule first and start to draw conclusions from the fact that tweety swims.

These strategies may help in getting reasonable behavior from a forward chaining system, but the most important thing is how we write the rules. They should be carefully constructed, with the preconditions specifying as precisely as possible when different rules should fire. Otherwise we will have little idea or control of what will happen. Sometimes special working memory elements are used to help to control the behavior of the system. For example, we might decide that there are certain basic stages of processing in doing some task, and certain rules should only be fired at a given stage - we could have a special working memory element (stage 1) and add (stage 1) to the preconditions of all the relevant rules, removing the working memory element when that stage was complete.

Choice between Forward and Backward Chaining

Forward chaining is often preferable in cases where there are many rules with the same conclusions. A well-known category of such rule systems are taxonomic hierarchies. E.g. the taxonomy of the animal kingdom includes such rules as:

```
animal(X) :- sponge(X).

animal(X) :- arthopod(X).

animal(X) :- vertebrate(X).

...

vertebrate(X) :- fish(X).

vertebrate(X) :- mammal(X)

...

mammal(X) :- carnivore(X)

...

carnivore(X) :- dog(X).

carnivore(X) :- cat(X).

...
```

(Skipped family and genus in the hierarchy.)

Now, suppose we have such a knowledge base of rules, we add the fact "dog(fido)" and we query whether "animal(fido)". In forward chaining, we will successively add "carnivore(fido)", "mammal(fido)", "vertebrate(fido)", and "animal(fido)". The query will then succeed immediately. The total work is proportional to the height of the hierarchy. By contast, if you use backward chaining, the query "~animal(fido)" will unify with the first rule above, and generate the subquery "~sponge(fido)", which will initiate a search for Fido through all the subdivisions of sponges, and so on. Ultimately, it searches the entire taxonomy of animals looking for Fido.

In some cases, it is desirable to *combine* forward and backward chaining. For example, suppose we augment the above animal with features of these various categories:

```
breathes(X) :- animal(X).

...

backbone(X) :- vertebrate(X).

has(X,brain) :- vertebrate(X).

...

furry(X) :- mammal(X).

warm_blooded(X) :- mammal(X).

...
```

If all these rules are implemented as forward chaining, then as soon as we state that Fido is a dog, we have to add all his known properties to Gamma; that he breathes, is warm-blooded, has a liver and kidney, and so on. The solution is to mark these property rules as backward chaining and mark the hierarchy rules as forward chaining. You then implement the knowledge base with both the forward chaining algorithm, restricted to rules marked as forward chaining, and backward chaining rules, restricted to rules marked as backward chaining. However, it is hard to guarantee that such a mixed inference system will be complete.

AND/OR Trees

We will next show the use of AND/OR trees for inferencing in Horn clause systems. The problem is, given a set of axioms in Horn clause form and a goal, show that the goal can be proven from the axioms.

An AND/OR tree is a tree whose internal nodes are labeled either "AND" or "OR". A *valuation* of an AND/OR tree is an assignment of "TRUE" or "FALSE" to each of the leaves. Given a tree T and a valuation over the leaves of T, the values of the internal nodes and of T are defined recursively in the obvious way:

- An OR node is TRUE if at least one of its children is TRUE.

- An AND node is TRUE if all of its children are TRUE.

The above is an *unconstrained* AND/OR tree. Also common are *constrained* AND/OR trees, in which the leaves labeled "TRUE" must satisfy some kind of constraint. A *solution* to a constrained AND/OR tree is a valuation that satisfies the constraint and gives the tree the value "TRUE".

An OR node is a goal to be proven. A goal G has one downward arc for each rule R whose head resolves with G. This leads to an AND node. The children in the AND node are the literals in the tail of R. Thus, a rule is satisfied if all its subgoals are satisfied (the AND node); a goal is satisfied if it is established by one of its rules (the OR node). The leaves are unit clauses, with no tail, labeled TRUE, and subgoals with no matching rules, labeled FALSE . The constraint is that the variable bindings must be consistent. The figure below show the AND/OR tree corresponding to the following Prolog rule set with the goal "common_ancestor(Z,edward,mary)"

```
Axioms:

/* Z is a common ancestor of X and Y */

R1: common_ancestor(Z,X,Y) :- ancestor(Z,X), ancestor(Z,Y).

/* Usual recursive definition of ancestor, going upward. */

R2: ancestor(A,A).

R3: ancestor(A,C) :- parent(B,C), ancestor(A,B).

/* Mini family tree */

R4: parent(catherine,mary).

R5: parent(henry,mary).

R6: parent(jane,edward).

R7: parent(henry,edward).
```

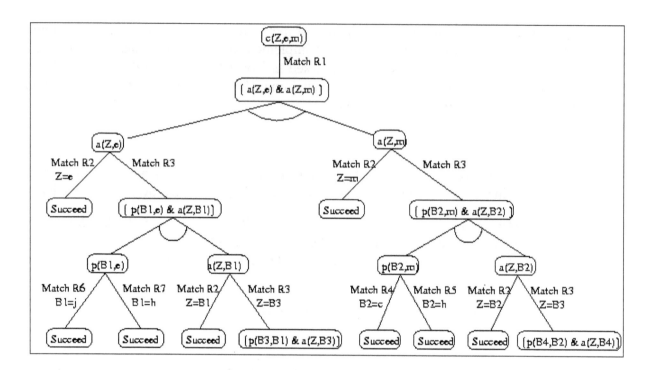

Programs in PROLOG

These minimal notes on Prolog show only some of its flavor.

Here are facts

```
plays(ann,fido).

friend(john,sam).
```

where ann, fido, john, and sam are individuals, and plays and friend are functors. And here is a rule

```
likes(X,Y) :- plays(X,Y),friend(X,Y).
```

It says that if X plays with Y and X is a friend of Y then X likes Y. Variables start with capital letters (If we are not interested in the value of a variable, we can just use _ (underscore)).

In a rule the left-hand side and the right-hand side are called respectively the head and the tail of the rule.

The prompt in prolog is

```
| ?-
```

You exit prolog with the statement

```
halt.
```

You can add rules and facts to the current session in a number of ways:

1. You can enter clauses from the terminal as follows:

2. | ?- consult(user).

3. | like(b,a).

4. | like(d,c).

5.   ^D

     which adds the two clauses to the working space.

6. Or you can read in a file:

7. | ?- consult('filename').

     which is added at the end of the working space.

8. Or you can assert a clause

9. | ?- assert(< clause >).

10.

Here are some confusing "equalities" in prolog:

predicate	relation	variable substitution	arithmetic computation
==	identical	no	no
=	unifiable	yes	no
=:=	same value	no	yes
is	is the value of	yes	yes

and some examples of their use:

     ?- X == Y.

     no

     ?- X + 1 == 2 + Y.

     no

     ?- X + 1 = 2 + Y.

     X = 2

     Y = 1

     ?- X + 1 = 1 + 2.

```
no
```

Example: Factorial1

```
factorial(0,1).

factorial(N,F) :- N>0, N1 is N-1, factorial(N1,F1),F is N*F1.
```

then

```
?- factorial(5,F).

F = 120

?- factorial(X,120).

instantiation error
```

Example: Factorial2: dropping N>0 from factorial1

```
factorial(0,1).

factorial(N,F) :- N1 is N-1, factorial(N1,F1),F is N*F1.
```

then

```
?- factorial(5,F).

F = 120; Here ";" asks for next value of F

keeps on going until stack overflow
```

Example: Factorial3: Changing order of terms

```
(1) factorial(0,1).

(2) factorial(N,F) :- factorial(N1,F1), N is N1+1, F is N*F1.
```

then

```
?- factorial(5,F).

F = 120

?- factorial(X,120).

X = 5;

integer overflow
```

Here is why factorial(X,120) returns 5. For brevity, instead of "factorial" we will write "f".

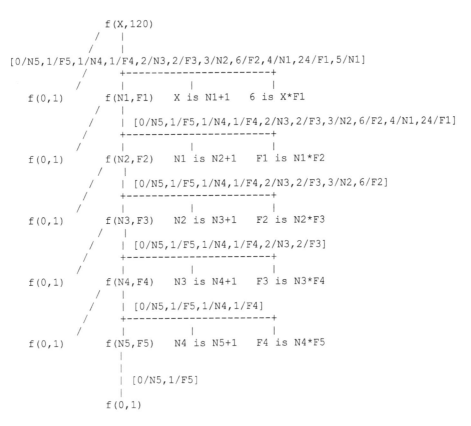

```
                            f(X,120)
                          /   |
                         /    |
          [0/N5,1/F5,1/N4,1/F4,2/N3,2/F3,3/N2,6/F2,4/N1,24/F1,5/N1]
                      /       +----------------------+
                     /        |            |         |
          f(0,1)        f(N1,F1)    X is N1+1    6 is X*F1
                      /   |
                     /    |  [0/N5,1/F5,1/N4,1/F4,2/N3,2/F3,3/N2,6/F2,4/N1,24/F1]
                    /     +----------------------+
                   /      |          |           |
          f(0,1)      f(N2,F2)    N1 is N2+1    F1 is N1*F2
                    /   |
                   /    |  [0/N5,1/F5,1/N4,1/F4,2/N3,2/F3,3/N2,6/F2]
                  /     +----------------------+
                 /      |          |         |
          f(0,1)      f(N3,F3)   N2 is N3+1    F2 is N2*F3
                    /   |
                   /    |  [0/N5,1/F5,1/N4,1/F4,2/N3,2/F3]
                  /     +----------------------+
                 /      |          |         |
          f(0,1)      f(N4,F4)   N3 is N4+1    F3 is N3*F4
                    /   |
                   /    |  [0/N5,1/F5,1/N4,1/F4]
                  /     +----------------------+
                 /      |          |         |
          f(0,1)      f(N5,F5)   N4 is N5+1    F4 is N4*F5
                        |
                        |
                        |  [0/N5,1/F5]
                        |
                     f(0,1)
```

In this diagram we see the substitutions computed. Much is not said in the diagram, for example why we abandon the unifications with the various f(0,1)s. [Let's say it for the second f(0,1) from the top: because it forces the substitution [0/N1,1/F1,1/X] and this cause 6 is X*F1 to fail.]

Lists

Lists are very much as in lisp. In place of Lisp's cons, in Prolog we use the "." or dot:

Dot Notation	List Notation	Lisp Notation
.(X,Y)	[X | Y]	(X . Y)
.(X, .(Y,Z))	[X,Y|Z]	(X (Y . Z))
.(X, .(Y, .(Z, [])))	[X,Y,Z]	(X Y Z)

Example: len

```
len([],0).

len([_|T], N) :- len(T,M), N is M+1.

?- len([a,b,c],X).

X = 3
```

```
?- len([a,b,c], 3).

yes

?- len([a,b,c], 5).

no
```

Example: member

member(X,Y) is inteded to mean X is one of the top level elements of the list Y.

```
member(X,[X|_]).

member(X,[_|T]) :- member(X,T).

?- member(X, [1,2,3,4,5]).

X=1;

X=2;

X=3;

X=4;

X=5;

no
```

Example: select

select(X,A,R) is intended to mean that X is a top level element of the list A and that R is what is left of A after taking X out of it.

```
select(H,[H|T],T).

select(X,[H|T],[H|T1]) :- select(X,T,T1).

?- select(X,[a,b,c],R).

X=a

R=[b,c];

X=b

R=[a,c];

X=c

R=[a,b];

No
```

The Cryptography Problem

Here is a problem:

```
    S  E  N  D  +
       M  O  R  E
    ---------
    M  O  N  E  Y
```

to be solved by mapping the letters into distinct digits and then doing regular arithmetic. We add variables to represent the various carries:

```
    C3 C2 C1
       S  E  N  D  +
          M  O  R  E
       ------------
       M  O  N  E  Y
```

We observe that carries can only be 0 or 1 and thus that M has to be 1. Then here is a solution:

```
solve([S,E,N,D], [M,O,R,E], [M,O,N,E,Y]) :-
  M=1,  L=[2,3,4,5,6,7,8,9],
  select(S,L,L1),  S>0,  (C3=0; C3=1),  ";" means OR
  O is S+M+C3-10*M, select(O, L1, L2),
  select(E,L2,L3),  (C2=0;C2=1),
  N is E+O+C2-10*C3, select(N,L3,L4),  (C1=0;C1=1),
  R is E+10*C2-(N+C1),  select(R,L4,L5),
  select(D,L5,L6),
  Y is D+E-10*C1, select(Y,L6,_).
?- solve([S,E,N,D], [M,O,R,E], [M,O,N,E,Y]).
  S=9
  E=5
  N=6
  D=7
  M=1
  O=0
  R=8
  Y=2;
  No
```

Expert Systems

An expert system is a computer program that contains some of the subject-specific knowledge of one or more human experts. An expert systems are meant to solve real problems which normally would require a specialized human expert (such as a doctor or a minerologist). Building an expert system therefore first involves extracting the relevant knowledge from the human expert. Such knowledge is often heuristic in nature, based on useful ``rules of thumb'' rather than absolute certainties. Extracting it from the expert in a way that can be used by a computer is generally a difficult task, requiring its own expertise. A *knowledge engineer* has the job of extracting this knowledge and building the expert system *knowledge base*.

A first attempt at building an expert system is unlikely to be very successful. This is partly because the expert generally finds it very difficult to express exactly what knowledge and rules they use to solve a problem. Much of it is almost subconscious, or appears so obvious they don't even bother mentioning it. *Knowledge acquisition* for expert systems is a big area of research, with a wide variety of techniques developed. However, generally it is important to develop an initial prototype based on information extracted by interviewing the expert, then iteratively refine it based on feedback both from the expert and from potential users of the expert system.

In order to do such iterative development from a prototype it is important that the expert system is written in a way that it can easily be inspected and modified. The system should be able to explain its reasoning (to expert, user and knowledge engineer) and answer questions about the solution process. Updating the system shouldn't involve rewriting a whole lot of code - just adding or deleting localized chunks of knowledge.

The most widely used knowledge representation scheme for expert systems is rules. Typically, the rules won't have certain conclusions - there will just be some degree of certainty that the conclusion will hold if the conditions hold. Statistical techniques are used to determine these certainties. Rule-based systems, with or without certainties, are generally easily modifiable and make it easy to provide reasonably helpful traces of the system's reasoning. These traces can be used in providing explanations of what it is doing.

Expert systems have been used to solve a wide range of problems in domains such as medicine, mathematics, engineering, geology, computer science, business, law, defence and education. Within each domain, they have been used to solve problems of different types. Types of problem involve *diagnosis* (e.g., of a system fault, disease or student error); *design* (of a computer systems, hotel etc); and *interpretation* (of, for example, geological data). The appropriate problem solving technique tends to depend more on the problem type than on the domain. Whole books have been written on how to choose your knowledge representation and reasoning methods given characteristics of your problem.

The following figure shows the most important modules that make up a rule-based expert system. The user interacts with the system through a *user interface* which may use menus, natural language or any other style of interaction). Then an *inference engine* is used to reason with both the *expert knowledge* (extracted from our friendly expert) and data specific to the particular problem being solved. The expert knowledge will typically be in the form of a set of IF-THEN rules. The *case specific data* includes both data provided by the user and partial conclusions (along with certainty measures) based on this data. In a simple forward chaining rule-based system the case specific data will be the elements in *working memory*.

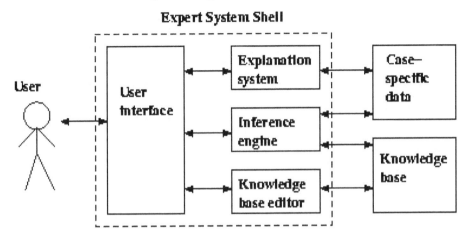

Almost all expert systems also have an *explanation subsystem,* which allows the program to explain its reasoning to the user. Some systems also have a *knowledge base editor* which help the expert or knowledge engineer to easily update and check the knowledge base.

One important feature of expert systems is the way they (usually) separate domain specific knowledge from more general purpose reasoning and representation techniques. The general purpose bit (in the dotted box in the figure) is referred to as an *expert system shell.* As we see in the figure, the shell will provide the inference engine (and knowledge representation scheme), a user interface, an explanation system and sometimes a knowledge base editor. Given a new kind of problem to solve (say, car design), we can usually find a shell that provides the right sort of support for that problem, so all we need to do is provide the expert knowledge. There are numerous commercial expert system shells, each one appropriate for a slightly different range of problems. (Expert systems work in industry includes both writing expert system shells and writing expert systems using shells.) Using shells to write expert systems generally greatly reduces the cost and time of development.

References

- Knublauch, Holger; Oberle, Daniel; Tetlow, Phil; Wallace, Evan (2006-03-09). "A Semantic Web Primer for Object-Oriented Software Developers". W3C. Retrieved 2008-07-30

- Brachman, Ron (1985). "Introduction". In Ronald Brachman and Hector J. Levesque. Readings in Knowledge Representation. Morgan Kaufmann. pp. XVI–XVII. ISBN 0-934613-01-X

- MacGregor, Robert (June 1991). "Using a description classifier to enhance knowledge representation". IEEE Expert. 6 (3): 41–46. doi:10.1109/64.87683. Retrieved 10 November 2013

- Bergmann, Merrie (2008). An introduction to many-valued and fuzzy logic: semantics, algebras, and derivation systems. Cambridge University Press. p. 114. ISBN 978-0-521-88128-9

- Berners-Lee, Tim; James Hendler; Ora Lassila (May 17, 2001). "The Semantic Web A new form of Web content that is meaningful to computers will unleash a revolution of new possibilities". Scientific American. 284: 34–43. doi:10.1038/scientificamerican0501-34. Archived from the original on April 24, 2013

- Smith, Brian C. (1985). "Prologue to Reflections and Semantics in a Procedural Language". In Ronald Brachman and Hector J. Levesque. Readings in Knowledge Representation. Morgan Kaufmann. pp. 31–40. ISBN 0-934613-01-X

- Zlatarva, Nellie (1992). "Truth Maintenance Systems and their Application for Verifying Expert System Knowledge Bases". Artificial Intelligence Review. 6: 67–110. doi:10.1007/bf00155580. Retrieved 25 December 2013

- Ebbinghaus, Heinz-Dieter; Flum, Jörg; and Thomas, Wolfgang (1994); Mathematical Logic, Undergraduate Texts in Mathematics, Berlin, DE/New York, NY: Springer-Verlag, Second Edition, ISBN 978-0-387-94258-2

- Macgregor, Robert (August 13, 1999). "Retrospective on Loom". isi.edu. Information Sciences Institute. Retrieved 10 December 2013

- Rautenberg, Wolfgang (2010), A Concise Introduction to Mathematical Logic (3rd ed.), New York, NY: Springer Science+Business Media, doi:10.1007/978-1-4419-1221-3, ISBN 978-1-4419-1220-6

- N. V. Murray (1982). "Completely Non-Clausal Theorem Proving". Artificial Intelligence. 18: 67–85. doi:10.1016/0004-3702(82)90011-x

Probabilistic Logic in Artificial Intelligence

Knowledge can be represented using logic in the form of facts and rules. However, in complex scenarios, it becomes difficult to do the same. In order to represent complexity, certainty factors are used. Uncertainty comes into play when real world situations are to be evaluated. This chapter elucidates the role and representation of uncertainty.

Probabilistic Logic

The aim of a probabilistic logic (also probability logic and probabilistic reasoning) is to combine the capacity of probability theory to handle uncertainty with the capacity of deductive logic to exploit structure of formal argument. The result is a richer and more expressive formalism with a broad range of possible application areas. Probabilistic logics attempt to find a natural extension of traditional logic truth tables: the results they define are derived through probabilistic expressions instead. A difficulty with probabilistic logics is that they tend to multiply the computational complexities of their probabilistic and logical components. Other difficulties include the possibility of counter-intuitive results, such as those of Dempster-Shafer theory. The need to deal with a broad variety of contexts and issues has led to many different proposals.

Historical Context

There are numerous proposals for probabilistic logics. Very roughly, they can be categorized into two different classes: those logics that attempt to make a probabilistic extension to logical entailment, such as Markov logic networks, and those that attempt to address the problems of uncertainty and lack of evidence (evidentiary logics).

That probability and uncertainty are not quite the same thing may be understood by noting that, despite the mathematization of probability in the Enlightenment, mathematical probability theory remains, to this very day, entirely unused in criminal courtrooms, when evaluating the "probability" of the guilt of a suspected criminal.

More precisely, in evidentiary logic, there is a need to distinguish the truth of a statement from the confidence in its truth: thus, being uncertain of a suspect's guilt is not the same as assigning a numerical probability to the commission of the crime. A single suspect may be guilty or not guilty, just as a coin may be flipped heads or tails. Given a large collection of suspects, a certain percentage may be guilty, just as the probability of flipping "heads" is one-half. However, it is incorrect to take this law of averages with regard to a single criminal (or single coin-flip): the criminal is no more "a little bit guilty" than a single coin flip is "a little bit heads and a little bit tails": we are merely uncertain as to which it is. Conflating probability and uncertainty may be acceptable when making scientific measurements of physical quantities, but it is an error, in the context of "common sense" reasoning and logic. Just as in courtroom reasoning, the goal of employing uncertain inference is to gather evidence to strengthen the confidence of a proposition, as opposed to performing some sort of probabilistic entailment.

Historically, attempts to quantify probabilistic reasoning date back to antiquity. There was a particularly strong interest starting in the 12th century, with the work of the Scholastics, with the invention of the half-proof (so that two half-proofs are sufficient to prove guilt), the elucidation of moral certainty (sufficient certainty to act upon, but short of absolute certainty), the development of Catholic probabilism (the idea that it is always safe to follow the established rules of doctrine or the opinion of experts, even when they are less probable), the case-based reasoning of casuistry, and the scandal of Laxism (whereby probabilism was used to give support to almost any statement at all, it being possible to find an expert opinion in support of almost any proposition.).

Modern Proposals

Below is a list of proposals for probabilistic and evidentiary extensions to classical and predicate logic.

- The term *"probabilistic logic"* was first used in a paper by Nils Nilsson published in 1986, where the truth values of sentences are probabilities. The proposed semantical generalization induces a probabilistic logical entailment, which reduces to ordinary logical entailment when the probabilities of all sentences are either 0 or 1. This generalization applies to any logical system for which the consistency of a finite set of sentences can be established.

- The central concept in the theory of subjective logic are *opinions* about some of the propositional variables involved in the given logical sentences. A binomial opinion applies to a single proposition and is represented as a 3-dimensional extension of a single probability value to express various degrees of ignorance about the truth of the proposition. For the computation of derived opinions based on a structure of argument opinions, the theory proposes respective operators for various logical connectives, such as e.g. multiplication (AND), comultiplication (OR), division (UN-AND) and co-division (UN-OR) of opinions as well as conditional deduction (MP) and abduction (MT).

- Approximate reasoning formalism proposed by fuzzy logic can be used to obtain a logic in which the models are the probability distributions and the theories are the lower envelopes. In such a logic the question of the consistency of the available information is strictly related with the one of the coherence of partial probabilistic assignment and therefore with Dutch book phenomenon.

- Markov logic networks implement a form of uncertain inference based on the maximum entropy principle—the idea that probabilities should be assigned in such a way as to maximize entropy, in analogy with the way that Markov chains assign probabilities to finite state machine transitions.

- Systems such as Pei Wang's Non-Axiomatic Reasoning System (NARS) or Ben Goertzel's Probabilistic Logic Networks (PLN) add an explicit confidence ranking, as well as a probability to atoms and sentences. The rules of deduction and induction incorporate this uncertainty, thus side-stepping difficulties in purely Bayesian approaches to logic (including

Markov logic), while also avoiding the paradoxes of Dempster-Shafer theory. The implementation of PLN attempts to use and generalize algorithms from logic programming, subject to these extensions.

- In the theory of probabilistic argumentation, probabilities are not directly attached to logical sentences. Instead it is assumed that a particular subset W of the variables V involved in the sentences defines a probability space over the corresponding sub-σ-algebra. This induces two distinct probability measures with respect to V, which are called *degree of support* and *degree of possibility*, respectively. Degrees of support can be regarded as non-additive *probabilities of provability*, which generalizes the concepts of ordinary logical entailment (for $V = \{\}$) and classical posterior probabilities (for $V = W$). Mathematically, this view is compatible with the Dempster-Shafer theory.

- The theory of evidential reasoning also defines non-additive *probabilities of probability* (or *epistemic probabilities*) as a general notion for both logical entailment (provability) and probability. The idea is to augment standard propositional logic by considering an epistemic operator K that represents the state of knowledge that a rational agent has about the world. Probabilities are then defined over the resulting *epistemic universe* K*p* of all propositional sentences *p*, and it is argued that this is the best information available to an analyst. From this view, Dempster-Shafer theory appears to be a generalized form of probabilistic reasoning.

Probabilistic Reasoning

Using logic to represent and reason we can represent knowledge about the world with facts and rules, like the following ones:

bird(tweety). fly(X) :- bird(X).

We can also use a theorem-prover to reason about the world and deduct new facts about the world, for e.g.,

?- fly(tweety). Yes

However, this often does not work outside of toy domains - non-tautologous certain rules are hard to find.

A way to handle knowledge representation in real problems is to extend logic by using certainty factors.

In other words, replace IF condition THEN fact with

IF condition with certainty x THEN fact with certainty $f(x)$

Unfortunately cannot really adapt logical inference to probabilistic inference, since the latter is not context-free.

Replacing rules with conditional probabilities makes inferencing simpler.

Replace

smoking -> lung cancer or

lotsofconditions, smoking -> lung cancer with

P(lung cancer | smoking) = 0.6

Uncertainty is represented explicitly and quantitatively within probability theory, a formalism that has been developed over centuries.

A probabilistic model describes the world in terms of a set S of possible states - the sample space. We don't know the true state of the world, so we (somehow) come up with a probability distribution over S which gives the probability of any state being the true one. The world usually described by a set of variables or attributes.

Consider the probabilistic model of a fictitious medical expert system. The 'world' is described by 8 binary valued variables:

Visit to Asia? A

Tuberculosis? T

Either tub. or lung cancer? E

Lung cancer? L

Smoking? S

Bronchitis? B

Dyspnoea? D

Positive X-ray? X

We have $2^8 = 256$ possible states or configurations and so 256 probabilities to find.

Review of Probability Theory

The primitives in probabilistic reasoning are *random variables*. Just like primitives in Propositional Logic are propositions. A random variable is not in fact a variable, but a function from a sample space S to another space, often the real numbers.

For example, let the random variable Sum (representing outcome of two die throws) be defined thus:

Sum(die1, die2) = die1 +die2

Each random variable has an associated probability distribution determined by the underlying distribution on the sample space

Continuing our example : P(Sum = 2) = 1/36,

P(Sum = 3) = 2/36, . . . , P(Sum = 12) = 1/36

Consdier the probabilistic model of the fictitious medical expert system mentioned before. The sample space is described by 8 binary valued variables.

Visit to Asia? A

Tuberculosis? T

Either tub. or lung cancer? E

Lung cancer? L

Smoking? S

Bronchitis? B

Dyspnoea? D

Positive X-ray? X

There are $2^8 = 256$ events in the sample space. Each event is determined by a joint instantiation of all of the variables.

$$S = \{(A = f, T = f, E = f, L = f, S = f, B = f, D = f, X = f),$$
$$(A = f, T = f, E = f, L = f, S = f, B = f, D = f, X = t), \ldots$$
$$\left(A = t, T = t, E = t, L = t, S = t, B = t, D = t, X = t\right)\}$$

Since S is defined in terms of joint instantations, any distribution defined on it is called a joint distribution. ll underlying distributions will be joint distributions in this module. The variables {A,T,E, L,S,B,D,X} are in fact random variables, which 'project' values.

$$L(A = f, T = f, E = f, L = f, S = f, B = f, D = f, X = f) = f$$
$$L(A = f, T = f, E = f, L = f, S = f, B = f, D = f, X = t) = f$$
$$L\left(A = t, T = t, E = t, L = t, S = t, B = t, D = t, X = t\right) = t$$

Each of the random variables {A,T,E,L,S,B,D,X} has its own distribution, determined by the underlying joint distribution. This is known as the margin distribution. For example, the distribution for L is denoted P(L), and this distribution is defined by the two probabilities P(L = f) and P(L = t). For example,

$$P\left(L = f\right)$$
$$= P(A = f, T = f, E = f, L = f, S = f, B = f, D = f, X = f)$$
$$+ P(A = f, T = f, E = f, L = f, S = f, B = f, D = t, X = t)$$
$$+ P(A = f, T = f, E = f, L = f, S = f, B = f, D = t, X = f)$$
$$\ldots$$
$$P\left(A = t, T = t, E = t, L = f, S = t, B = t, D = t, X = t\right)$$

P(L) is an example of a marginal distribution.

Here's a joint distribution over two binary value variables A and B

	A=0	A=1
B=0	0.2	0.3
B=1	0.4	0.1

We get the marginal distribution over B by simply adding up the different possible values of A for any value of B (and put the result in the "margin").

	A=0	A=1	
B=0	0.2	0.3	0.5 (= 0.2+0.3)
B=1	0.4	0.1	0.5 (=0.4 + 0.1)

In general, given a joint distribution over a set of variables, we can get the marginal distribution over a subset by simply summing out those variables not in the subset.

In the medical expert system case, we can get the marginal distribution over, say, A,D by simply summing out the other variables:

$$P(A,D) = \sum_{T,E,L,S,B,X} P(A,T,E,L,S,B,D,X)$$

However, computing marginals is not an easy task always. For example,

$$P(A = t, D = f)$$
$$= P(A = t, T = f, E = f, L = f, S = f, B = f, D = f, X = f)$$
$$+ P(A = t, T = f, E = f, L = f, S = f, B = f, D = f, X = t)$$
$$+ P(A = t, T = f, E = f, L = f, S = f, B = t, D = f, X = f)$$
$$+ P(A = t, T = f, E = f, L = f, S = f, B = t, D = f, X = t)$$
$$\cdots$$
$$P(A = t, T = t, E = t, L = t, S = t, B = t, D = f, X = t)$$

This has 64 summands. Each of whose value needs to be estimated from empirical data. For the estimates to be of good quality, each of the instances that appear in the summands should appear sufficiently large number of times in the empirical data. Often such a large amount of data is not available.

However, computation can be simplified for certain special but common conditions. This is the condition of *independence* of variables.

Two random variables A and B are independent if

P(A,B) = P(A)P(B)

i.e. can get the joint from the marginals

This is quite a strong statement: It means for any value x of A and any value y of B

$$P(A = x, B = y) \ = \ P(A \ = x)P(B = y)$$

Note that the independence of two random variables is a property of a the underlying

$$P(A, B) = P(A)P(B) \quad P'(A, B) \neq P'(A)P'(B)$$

probability distribution. We can have

Conditional probability is defined as:

$$P(A|B) \overset{\text{def}}{=} \frac{P(A, B)}{P(B)}$$

It means for any value x of A and any value y of B

$$P(A = x|B = y) = \frac{P(A = x, B = y)}{P(B = y)}$$

If A and B are independent then

$$P(A|B) = P(A)$$

Conditional probabilities can represent causal relationships in both directions.

From cause to (probable) effects

$$Car\_start = f \leftarrow Cold\_battery = t$$
$$P(Car\_start = f \mid Cold\_battery = t) = 0.8$$

From effect to (probable) cause

$$Cold\_battery = t \leftarrow Car\_start = f$$
$$P(Cold\_battery = t \mid Car\_start = f) = 0.7$$

Probabilistic Inference

Two rules in probability theory are important for inferencing, namely, the product rule and the Bayes' rule.

Here

Product rule:

$$P(A, B|C) = P(A|B, C)P(B|C)$$
$$= P(B|A, C)P(A|C)$$

Bayes' rule:

$$P(A|B,C) = \frac{P(B|A,C)P(A|C)}{P(B|C)}$$

Used in Bayesian statistics:

$$P(\text{Model Data}) = \frac{P(\text{Model})P(\text{Data}|\text{Model})}{P(\text{Data})}$$

Here is a simple example, of application of Bayes' rule.

Suppose you have been tested positive for a disease; what is the probability that you actually have the disease?

It depends on the accuracy and sensitivity of the test, and on the background (*prior*) probability of the disease.

Let P(Test=+ve | Disease=true) = 0.95, so the false negative rate, P(Test=-ve | Disease=true), is 5%.

Let P(Test=+ve | Disease=false) = 0.05, so the false positive rate is also 5%. Suppose the disease is rare: P(Disease=true) = 0.01 (1%).

Let D denote Disease and "T=+ve" denote the positive Tes.

Then,

```
                              P(T=+ve|D=true) * P(D=true)
P(D=true|T=+ve) = ----------------------------------------------------------------
                   P(T=+ve|D=true) * P(D=true)+ P(T=+ve|D=false) * P(D=false)


                      0.95 * 0.01
               = --------------------------------   =    0.161
                   0.95*0.01 + 0.05*0.99
```

So the probability of having the disease given that you tested positive is just 16%. This seems too low, but here is an intuitive argument to support it. Of 100 people, we expect only 1 to have the disease, but we expect about 5% of those (5 people) to test positive. So of the 6 people who test positive, we only expect 1 of them to actually have the disease; and indeed 1/6 is approximately 0.16.

In other words, the reason the number is so small is that you believed that this is a rare disease; the test has made it 16 times more likely you have the disease, but it is still unlikely in absolute terms. If you want to be "objective", you can set the prior to uniform (i.e. effectively ignore the prior), and then get

```
                            P(T=+ve|D=true) * P(D=true)
P(D=true|T=+ve) = -----------------------------------------------------------
                                  P(T=+ve)


                   0.95 * 0.5            0.475
               = --------------------- = ------- = 0.95
                  0.95*0.5 + 0.05*0.5     0.5
```

This, of course, is just the true positive rate of the test. However, this conclusion relies on your belief that, if you did not conduct the test, half the people in the world have the disease, which does not seem reasonable.

A better approach is to use a plausible prior (eg P(D=true)=0.01), but then conduct multiple independent tests; if they all show up positive, then the posterior will increase. For example, if we conduct two (conditionally independent) tests T1, T2 with the same reliability, and they are both positive, we get

$$P(D=true|T1=+ve,T2=+ve) = \frac{P(T1=+ve|D=true) * P(T2=+ve|D=true) * P(D=true)}{P(T1=+ve, T2=+ve)}$$

$$= \frac{0.95 * 0.95 * 0.01}{0.95*0.95*0.01 + 0.05*0.05*0.99} = \frac{0.009}{0.0115} = 0.7826$$

The assumption that the pieces of evidence are conditionally independent is called the naive Bayes assumption. This model has been successfully used for mainly application including classifying email as spam (D=true) or not (D=false) given the presence of various key words (Ti=+ve if word i is in the text, else Ti=-ve). It is clear that the words are not independent, even conditioned on spam/not-spam, but the model works surprisingly well nonetheless.

In many problems, complete independence of variables do not exist. Though many of them are conditionally independent.

X and Y are conditionally independent given Z iff

$$P(X,Y \mid Z) = P(X \mid Z)P(Y \mid Z)$$

In full: X and Y are conditionally independent given Z iff for any instantiation x, y, z of X, Y,Z we have

$$P(X = x, Y = y \mid Z = z)$$
$$= P(X = x, Z = z)P(Y = y \mid Z = z)$$

An example of conditional independence:

Consider the following three Boolean random variables:

LeaveBy8, GetTrain, OnTime

Suppose we can assume that:

P(*OnTime* | *GetTrain, LeaveBy8*) = P(*OnTime* | *GetTrain*)

but NOT   P(*OnTime* | *LeaveBy8*) = P(*OnTime*)

Then, *OnTime* is dependent on *LeaveBy8*, but *independent of LeaveBy8 given GetTrain*.

We can represent P(*OnTime* | *GetTrain, LeaveBy8*) = P(*OnTime* | *GetTrain*)

graphically by: *LeaveBy8 -> GetTrain -> OnTime*

Bayesian Network

Representation and Syntax

Bayes nets (BN) (also referred to as Probabilistic Graphical Models and Bayesian Belief Networks) are directed acyclic graphs (DAGs) where each node represents a random variable. The intuitive meaning of an arrow from a parent to a child is that the parent directly influences the child. These influences are quantified by conditional probabilities.

BNs are graphical representations of joint distributions. The BN for the medical expert system mentioned previously represents a joint distribution over 8 binary random variables {A,T,E,L,S,B,D,X}.

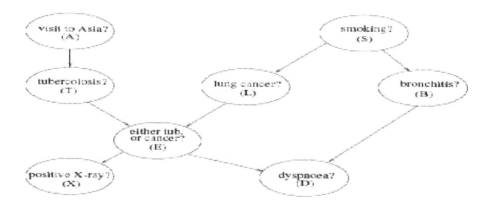

Conditional Probability Tables

Each node in a Bayesian net has an associated conditional probability table or CPT. (Assume all random variables have only a finite number of possible values). This gives the probability values for the random variable at the node conditional on values for its parents. Here is a part of one of the CPTs from the medical expert system network.

$$P(D = t \mid E = t, B = t) = 0.9 \quad P(D = t \mid E = t, B = f) = 0.7$$
$$P(D = t \mid E = f, B = t) = 0.8 \quad P(D = t \mid E = f, B = f) = 0.1$$

If a node has no parents, then the CPT reduces to a table giving the marginal distribution on that random variable.

$$P(A = f) = 0.1$$
$$P(A = f) = 0.9$$

Consider another example, in which all nodes are binary, i.e., have two possible values, which we will denote by T (true) and F (false).

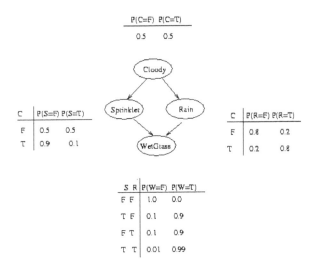

We see that the event "grass is wet" (W=true) has two possible causes: either the water sprinker is on (S=true) or it is raining (R=true). The strength of this relationship is shown in the table. For example, we see that Pr(W=true | S=true, R=false) = 0.9 (second row), and hence, Pr(W=false | S=true, R=false) = 1 - 0.9 = 0.1, since each row must sum to one. Since the C node has no parents, its CPT specifies the prior probability that it is cloudy (in this case, 0.5). (Think of C as representing the season: if it is a cloudy season, it is less likely that the sprinkler is on and more likely that the rain is on.)

Bayesian Network

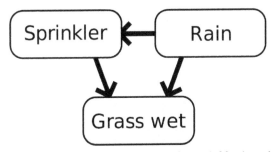

A simple Bayesian network. Rain influences whether the sprinkler is activated, and both rain and the sprinkler influence whether the grass is wet.

A Bayesian network, Bayes network, belief network, Bayes(ian) model or probabilistic directed acyclic graphical model is a probabilistic graphical model (a type of statistical model) that represents a set of random variables and their conditional dependencies via a directed acyclic graph (DAG). For example, a Bayesian network could represent the probabilistic relationships between diseases and symptoms. Given symptoms, the network can be used to compute the probabilities of the presence of various diseases.

Formally, Bayesian networks are DAGs whose nodes represent random variables in the Bayesian sense: they may be observable quantities, latent variables, unknown parameters or hypotheses. Edges represent conditional dependencies; nodes that are not connected (there is no path from one of the variables to the other in the Bayesian network) represent variables that are conditionally independent of each other. Each node is associated with a probability function that takes, as input, a particular set of values for the node's parent variables, and gives (as output) the probability (or probability distribution, if applicable) of the variable represented by the node. For example, if m parent nodes represent m Boolean variables then the probability function could be represented by a table of 2^m entries, one entry for each of the 2^m possible combinations of its parents being true or false. Similar ideas may be applied to undirected, and possibly cyclic, graphs; such as Markov networks.

Efficient algorithms exist that perform inference and learning in Bayesian networks. Bayesian networks that model sequences of variables (*e.g.* speech signals or protein sequences) are called dynamic Bayesian networks. Generalizations of Bayesian networks that can represent and solve decision problems under uncertainty are called influence diagrams.

Example

Suppose that there are two events which could cause grass to be wet: either the sprinkler is on or it's raining. Also, suppose that the rain has a direct effect on the use of the sprinkler (namely that when it rains, the sprinkler is usually not turned on). Then the situation can be modeled with a Bayesian network. All three variables have two possible values, T (for true) and F (for false).

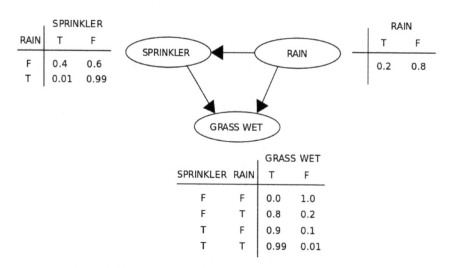

SPRINKLER		
RAIN	T	F
F	0.4	0.6
T	0.01	0.99

RAIN	
T	F
0.2	0.8

SPRINKLER	RAIN	GRASS WET T	GRASS WET F
F	F	0.0	1.0
F	T	0.8	0.2
T	F	0.9	0.1
T	T	0.99	0.01

A simple Bayesian network with conditional probability tables

The joint probability function is:

$$\Pr(G,S,R) = \Pr(G\,|\,S,R)\Pr(S\,|\,R)\Pr(R)$$

where the names of the variables have been abbreviated to G = *Grass wet (yes/no)*, S = *Sprinkler turned on (yes/no)*, and R = *Raining (yes/no)*.

The model can answer questions like "What is the probability that it is raining, given the grass is wet?" by using the conditional probability formula and summing over all nuisance variables:

$$\Pr(R=T\,|\,G=T) = \frac{\Pr(G=T,R=T)}{\Pr(G=T)} = \frac{\sum_{S\in\{T,F\}}\Pr(G=T,S,R=T)}{\sum_{S,R\in\{T,F\}}\Pr(G=T,S,R)}$$

Using the expansion for the joint probability function $\Pr(G, S, R)$ and the conditional probabilities from the conditional probability tables (CPTs) stated in the diagram, one can evaluate each term in the sums in the numerator and denominator. For example,

$$\Pr(G = T, S = T, R = T) = \Pr(G = T \mid S = T, R = T)\Pr(S = T \mid R = T)\Pr(R = T):$$

$$= 0.99 \times 0.01 \times 0.2 = 0.00198.$$

Then the numerical results (subscripted by the associated variable values) are

$$\Pr(R = T \mid G = T) = \frac{0.00198_{TTT} + 0.1584_{TFT}}{0.00198_{TTT} + 0.288_{TTF} + 0.1584_{TFT} + 0.0_{TFF}} = \frac{891}{2491} \approx 35.77\%.$$

If, on the other hand, we wish to answer an interventional question: "What is the probability that it would rain, given that we wet the grass?" the answer would be governed by the post-intervention joint distribution function $\Pr(S, R \mid \mathrm{do}(G = T)) = \Pr(S \mid R)P(R)$ obtained by removing the factor $\Pr(G \mid S, R)$ from the pre-intervention distribution. As expected, the probability of rain is unaffected by the action: $\Pr(R \mid \mathrm{do}(G = T)) = \Pr(R)$..

If, moreover, we wish to predict the impact of turning the sprinkler on, we have

$$\Pr(R, G \mid \mathrm{do}(S = T)) = \Pr(R)\Pr(G \mid R, S = T)$$

with the term $\Pr(S = T \mid R)$ removed, showing that the action has an effect on the grass but not on the rain.

These predictions may not be feasible when some of the variables are unobserved, as in most policy evaluation problems. The effect of the action $\mathrm{do}(x)$ can still be predicted, however, whenever a criterion called "back-door" is satisfied. It states that, if a set Z of nodes can be observed that d-separates (or blocks) all back-door paths from X to Y then $\Pr(Y, Z \mid \mathrm{do}(x)) = \Pr(Y, Z, X = x) / \Pr(X = x \mid Z)$. A back-door path is one that ends with an arrow into X. Sets that satisfy the back-door criterion are called "sufficient" or "admissible." For example, the set $Z = R$ is admissible for predicting the effect of $S = T$ on G, because R d-separate the (only) back-door path $S \leftarrow R \rightarrow G$. However, if S is not observed, there is no other set that d-separates this path and the effect of turning the sprinkler on $(S = T)$ on the grass (G) cannot be predicted from passive observations. We then say that $P(G \mid \mathrm{do}(S = T))$ is not "identified." This reflects the fact that, lacking interventional data, we cannot determine if the observed dependence between S and G is due to a causal connection or is spurious (apparent dependence arising from a common cause, R).

To determine whether a causal relation is identified from an arbitrary Bayesian network with unobserved variables, one can use the three rules of "do-calculus" and test whether all do terms can be removed from the expression of that relation, thus confirming that the desired quantity is estimable from frequency data.

Using a Bayesian network can save considerable amounts of memory, if the dependencies in the joint distribution are sparse. For example, a naive way of storing the conditional probabilities of 10 two-valued variables as a table requires storage space for $2^{10} = 1024$ values. If the local distributions of no variable depends on more than three parent variables, the Bayesian network representation only needs to store at most $10 \cdot 2 = 80$ values.

One advantage of Bayesian networks is that it is intuitively easier for a human to understand (a sparse set of) direct dependencies and local distributions than complete joint distributions.

Inference and Learning

There are three main inference tasks for Bayesian networks.

Inferring Unobserved Variables

Because a Bayesian network is a complete model for the variables and their relationships, it can be used to answer probabilistic queries about them. For example, the network can be used to find out updated knowledge of the state of a subset of variables when other variables (the *evidence* variables) are observed. This process of computing the *posterior* distribution of variables given evidence is called probabilistic inference. The posterior gives a universal sufficient statistic for detection applications, when one wants to choose values for the variable subset which minimize some expected loss function, for instance the probability of decision error. A Bayesian network can thus be considered a mechanism for automatically applying Bayes' theorem to complex problems.

The most common exact inference methods are: variable elimination, which eliminates (by integration or summation) the non-observed non-query variables one by one by distributing the sum over the product; clique tree propagation, which caches the computation so that many variables can be queried at one time and new evidence can be propagated quickly; and recursive conditioning and AND/OR search, which allow for a space-time tradeoff and match the efficiency of variable elimination when enough space is used. All of these methods have complexity that is exponential in the network's treewidth. The most common approximate inference algorithms are importance sampling, stochastic MCMC simulation, mini-bucket elimination, loopy belief propagation, generalized belief propagation, and variational methods.

Parameter Learning

In order to fully specify the Bayesian network and thus fully represent the joint probability distribution, it is necessary to specify for each node X the probability distribution for X conditional upon X's parents. The distribution of X conditional upon its parents may have any form. It is common to work with discrete or Gaussian distributions since that simplifies calculations. Sometimes only constraints on a distribution are known; one can then use the principle of maximum entropy to determine a single distribution, the one with the greatest entropy given the constraints. (Analogously, in the specific context of a dynamic Bayesian network, one commonly specifies the conditional distribution for the hidden state's temporal evolution to maximize the entropy rate of the implied stochastic process.)

Often these conditional distributions include parameters which are unknown and must be estimated from data, sometimes using the maximum likelihood approach. Direct maximization of the likelihood (or of the posterior probability) is often complex when there are unobserved variables. A classical approach to this problem is the expectation-maximization algorithm which alternates computing expected values of the unobserved variables conditional on observed data, with maximizing the complete likelihood (or posterior) assuming that previously computed expected values are correct. Under mild regularity conditions this process converges on maximum likelihood (or maximum posterior) values for parameters.

A more fully Bayesian approach to parameters is to treat parameters as additional unobserved variables and to compute a full posterior distribution over all nodes conditional upon observed data, then to integrate out the parameters. This approach can be expensive and lead to large dimension models, so in practice classical parameter-setting approaches are more common.

Structure Learning

In the simplest case, a Bayesian network is specified by an expert and is then used to perform inference. In other applications the task of defining the network is too complex for humans. In this case the network structure and the parameters of the local distributions must be learned from data.

Automatically learning the graph structure of a Bayesian network (BN) is a challenge pursued within machine learning. The basic idea goes back to a recovery algorithm developed by Rebane and Pearl (1987) and rests on the distinction between the three possible types of adjacent triplets allowed in a directed acyclic graph (DAG):

1. $X \to Y \to Z$

2. $X \leftarrow Y \to Z$

3. $X \to Y \leftarrow Z$

Type 1 and type 2 represent the same dependencies (X and Z are independent given Y) and are, therefore, indistinguishable. Type 3, however, can be uniquely identified, since X and Z are marginally independent and all other pairs are dependent. Thus, while the *skeletons* (the graphs stripped of arrows) of these three triplets are identical, the directionality of the arrows is partially identifiable. The same distinction applies when X and Z have common parents, except that one must first condition on those parents. Algorithms have been developed to systematically determine the skeleton of the underlying graph and, then, orient all arrows whose directionality is dictated by the conditional independencies observed.

An alternative method of structural learning uses optimization based search. It requires a scoring function and a search strategy. A common scoring function is posterior probability of the structure given the training data, like the BIC or the BDeu. The time requirement of an exhaustive search returning a structure that maximizes the score is superexponential in the number of variables. A local search strategy makes incremental changes aimed at improving the score of the structure. A global search algorithm like Markov chain Monte Carlo can avoid getting trapped in local minima. Friedman et al. discuss using mutual information between variables and finding a structure that maximizes this. They do this by restricting the parent candidate set to k nodes and exhaustively searching therein.

A particularly fast method for exact BN learning is to cast the problem as an optimization problem, and solve it using integer programming. Acyclicity constraints are added to the integer program (IP) during solving in the form of cutting planes. Such method can handle problems with up to 100 variables.

In order to deal with problems with thousands of variables, it is necessary to use a different approach. One is to first sample one ordering, and then find the optimal BN structure with respect to that ordering. This implies working on the search space of the possible orderings, which is convenient as it is smaller than the space of network structures. Multiple orderings are then sampled and evaluated. This method has been proven to be the best available in literature when the number of variables is huge.

Another method consists of focusing on the sub-class of decomposable models, for which the MLE have a closed form. It is then possible to discover a consistent structure for hundreds of variables.

As previously noted, learning Bayesian networks with bounded treewidth is necessary to allow exact tractable inference, since the worst-case inference complexity is exponential in the treewidth k (under the exponential time hypothesis). Yet, being a global property of the graph, it considerably increases the difficulty of the learning process. In this context it is possible to use the concept of K-tree for effective learning.

Statistical Introduction

Given data x and parameter θ, a simple Bayesian analysis starts with a prior probability (*prior*) $p(\theta)$ and likelihood $p(x \mid \theta)$ to compute a posterior probability $p(\theta \mid x) \propto p(x \mid \theta) p(\theta)$.

Often the prior on θ depends in turn on other parameters φ that are not mentioned in the likelihood. So, the prior $p(\theta)$ must be replaced by a likelihood $p(\theta \mid \varphi)$, and a prior $p(\varphi)$ on the newly introduced parameters φ is required, resulting in a posterior probability

$$p(\theta, \varphi \mid x) \propto p(x \mid \theta) p(\theta \mid \varphi) p(\varphi).$$

This is the simplest example of a *hierarchical Bayes model*.

The process may be repeated; for example, the parameters φ may depend in turn on additional parameters ψ, which will require their own prior. Eventually the process must terminate, with priors that do not depend on any other unmentioned parameters.

Introductory Examples

Suppose we have measured the quantities $x_1, \ldots, x_n$ each with normally distributed errors of known standard deviation σ,

$$x_i \sim N(\theta_i, \sigma^2)$$

Suppose we are interested in estimating the θ_i. An approach would be to estimate the θ_i using a maximum likelihood approach; since the observations are independent, the likelihood factorizes and the maximum likelihood estimate is simply

$$\theta_i = x_i$$

However, if the quantities are related, so that for example we may think that the individual θ_i have themselves been drawn from an underlying distribution, then this relationship destroys the independence and suggests a more complex model, e.g.,

$$x_i \sim N(\theta_i, \sigma^2),$$

$$\theta_i \sim N(\varphi, \tau^2)$$

with improper priors $\varphi \sim$ flat, $\tau \sim$ flat $\in (0, \infty)$. When $n \geq 3$, this is an *identified model* (i.e. there exists a unique solution for the model's parameters), and the posterior distributions of the individual θ_i will tend to move, or *shrink* away from the maximum likelihood estimates towards their common mean. This *shrinkage* is a typical behavior in hierarchical Bayes models.

Restrictions on Priors

Some care is needed when choosing priors in a hierarchical model, particularly on scale variables at higher levels of the hierarchy such as the variable τ in the example. The usual priors such as the Jeffreys prior often do not work, because the posterior distribution will be improper (not normalizable), and estimates made by minimizing the expected loss will be inadmissible.

Definitions and Concepts

There are several equivalent definitions of a Bayesian network. For all the following, let $G = (V, E)$ be a directed acyclic graph (or DAG), and let $X = (X_v)_{v \in V}$ be a set of random variables indexed by V.

Factorization Definition

X is a Bayesian network with respect to G if its joint probability density function (with respect to a product measure) can be written as a product of the individual density functions, conditional on their parent variables:

$$p(x) = \prod_{v \in V} p\left(x_v \mid x_{\mathrm{pa}(v)}\right)$$

where pa(v) is the set of parents of v (i.e. those vertices pointing directly to v via a single edge).

For any set of random variables, the probability of any member of a joint distribution can be calculated from conditional probabilities using the chain rule (given a topological ordering of X) as follows:

$$P(X_1 = x_1, \ldots, X_n = x_n) = \prod_{v=1}^{n} P\left(X_v = x_v \mid X_{v+1} = x_{v+1}, \ldots, X_n = x_n\right)$$

Compare this with the definition above, which can be written as:

$$P(X_1 = x_1, \ldots, X_n = x_n) = \prod_{v=1}^{n} P(X_v = x_v \mid X_j = x_j \text{ for each } X_j \text{ which is a parent of } X_v)$$

The difference between the two expressions is the conditional independence of the variables from any of their non-descendants, given the values of their parent variables.

Local Markov Property

X is a Bayesian network with respect to G if it satisfies the *local Markov property*: each variable is conditionally independent of its non-descendants given its parent variables:

$$X_v \perp\!\!\!\perp X_{V \setminus de(v)} \mid X_{pa(v)} \quad \text{for all } v \in V$$

where $de(v)$ is the set of descendants and $V \setminus de(v)$ is the set of non-descendants of v.

This can also be expressed in terms similar to the first definition, as

$P(X_v = x_v \mid X_i = x_i$ for each X_i which is not a descendant of $X_v) = P(X_v = x_v \mid X_j = x_j$ for each X_j which is a parent of $X_v)$

Note that the set of parents is a subset of the set of non-descendants because the graph is acyclic.

Developing Bayesian Networks

To develop a Bayesian network, we often first develop a DAG G such that we believe X satisfies the local Markov property with respect to G. Sometimes this is done by creating a causal DAG. We then ascertain the conditional probability distributions of each variable given its parents in G. In many cases, in particular in the case where the variables are discrete, if we define the joint distribution of X to be the product of these conditional distributions, then X is a Bayesian network with respect to G.

Markov Blanket

The Markov blanket of a node is the set of nodes consisting of its parents, its children, and any other parents of its children. The Markov blanket renders the node independent of the rest of the network; the joint distribution of the variables in the Markov blanket of a node is sufficient knowledge for calculating the distribution of the node. X is a Bayesian network with respect to G if every node is conditionally independent of all other nodes in the network, given its Markov blanket.

d-separation

This definition can be made more general by defining the "d"-separation of two nodes, where d stands for directional. Let P be a trail from node u to v. A trail is a loop-free, undirected path between two nodes (i.e. the direction of edges is ignored for constructing the path), in which edges may have any direction. Then P is said to be *d*-separated by a set of nodes Z if any of the following conditions holds:

1. *P* contains a directed *chain*, $u \ldots \leftarrow m \leftarrow \ldots v$ or $u \ldots \rightarrow m \rightarrow \ldots v$, such that the middle node m is in Z,

2. *P* contains a *fork*, $u \ldots \leftarrow m \rightarrow \ldots v$, such that the middle node m is in Z, or

3. *P* contains an *inverted fork* (or *collider*), $u \ldots \rightarrow m \leftarrow \ldots v$, such that the middle node m is not in Z and no descendant of m is in Z.

Thus u and v are said to be d-separated by Z if all trails between them are d-separated. If u and v are not d-separated, they are called d-connected.

X is a Bayesian network with respect to G if, for any two nodes u, v:

$$X_u \perp\!\!\!\perp X_v \mid X_Z$$

where Z is a set which d-separates u and v. (The Markov blanket is the minimal set of nodes which d-separates node v from all other nodes.)

Hierarchical Models

The term *hierarchical model* is sometimes considered a particular type of Bayesian network, but has no formal definition. Sometimes the term is reserved for models with three or more levels of random variables; other times, it is reserved for models with latent variables. In general, however, any moderately complex Bayesian network is usually termed "hierarchical".

Causal Networks

Although Bayesian networks are often used to represent causal relationships, this need not be the case: a directed edge from u to v does not require that X_v is causally dependent on X_u. This is demonstrated by the fact that Bayesian networks on the graphs:

$$a \rightarrow b \rightarrow c \quad \text{and} \quad a \leftarrow b \leftarrow c$$

are equivalent: that is they impose exactly the same conditional independence requirements.

A causal network is a Bayesian network with an explicit requirement that the relationships be causal. The additional semantics of the causal networks specify that if a node X is actively caused to be in a given state x (an action written as do($X = x$)), then the probability density function changes to the one of the network obtained by cutting the links from the parents of X to X, and setting X to the caused value x. Using these semantics, one can predict the impact of external interventions from data obtained prior to intervention.

Inference Complexity and Approximation Algorithms

In 1990 while working at Stanford University on large bioinformatic applications, Greg Cooper proved that exact inference in Bayesian networks is NP-hard. This result prompted a surge in research on approximation algorithms with the aim of developing a tractable approximation to probabilistic inference. In 1993, Paul Dagum and Michael Luby proved two surprising results on the complexity of approximation of probabilistic inference in Bayesian networks. First, they proved that there is no tractable *deterministic algorithm* that can approximate probabilistic inference to within an absolute error $\varepsilon < 1/2$. Second, they proved that there is no tractable *randomized algorithm* that can approximate probabilistic inference to within an absolute error $\varepsilon < 1/2$ with confidence probability greater than $1/2$.

At about the same time, Dan Roth proved that exact inference in Bayesian networks is in fact #P-complete (and thus as hard as counting the number of satisfying assignments of a CNF formula) and that approximate inference, even for Bayesian networks with restricted architecture, is NP-hard.

In practical terms, these complexity results suggested that while Bayesian networks were rich representations for AI and machine learning applications, their use in large real-world applications would need to be tempered by either topological structural constraints, such as naïve Bayes networks, or by restrictions on the conditional probabilities. The *bounded variance algorithm* was the first provable fast approximation algorithm to efficiently approximate probabilistic inference in Bayesian networks with guarantees on the error approximation. This powerful algorithm required the minor restriction on the conditional probabilities of the Bayesian network to be bounded away from zero and one by $1/p(n)$ where $p(n)$ was any polynomial on the number of nodes in the network n.

Applications

Bayesian networks are used for modelling beliefs in computational biology and bioinformatics (gene regulatory networks, protein structure, gene expression analysis, learning epistasis from GWAS data sets) medicine, biomonitoring, document classification, information retrieval, semantic search, image processing, data fusion, decision support systems, engineering, sports betting, gaming, law, study design and risk analysis. There are texts applying Bayesian networks to bioinformatics and financial and marketing informatics.The publications by Anderson and Vastag (2004), Lauría and Duchessi (2006), Gupta and Kim (2008), Lee et al. (2011) , Cardenas, Voordijk, and Dewulf (2017) have shown applications of Bayesian Networks to very disparate fields.

Software

- libDAI A free and open source C++ library for Discrete Approximate Inference in graphical models. libDAI supports such inference methods as exact inference by brute force enumeration, exact inference by junction-tree methods, Mean Field, Loopy Belief Propagation, Gibbs sampler, Conditioned Belief Propagation and some others.

- Mocapy++ A Dynamic Bayesian Network toolkit, implemented in C++. It supports discrete, multinomial, Gaussian, Kent, Von Mises and Poisson nodes. Inference and learning is done by Gibbs sampling/Stochastic-EM.

- WinBUGS One of the first computational implementations of MCMC samplers. No longer maintained and not recommended for active use.

- OpenBUGS (website), further (open source) development of WinBUGS.

- Just another Gibbs sampler (JAGS) (website) Another open source alternative to WinBUGS. Uses Gibbs sampling.

- Stan (software) (website) — Stan is an open-source package for obtaining Bayesian inference using the No-U-Turn sampler, a variant of Hamiltonian Monte Carlo. It's somewhat like BUGS, but with a different language for expressing models and a different sampler for sampling from their posteriors. RStan is the R interface to Stan. It is maintained by Andrew Gelman and colleagues.

- Direct Graphical Models (DGM) is an open source C++ library, implementing various tasks in probabilistic graphical models with pairwise dependencies.

- OpenMarkov, open source software and API implemented in Java

- Graphical Models Toolkit (GMTK) — GMTK is an open source, publicly available toolkit for rapidly prototyping statistical models using dynamic graphical models (DGMs) and dynamic Bayesian networks (DBNs). GMTK can be used for applications and research in speech and language processing, bioinformatics, activity recognition, and any time series application.

- PyMC — PyMC is a python module that implements Bayesian statistical models and fitting algorithms, including Markov chain Monte Carlo. Its flexibility and extensibility make it applicable to a large suite of problems. Along with core sampling functionality, PyMC includes methods for summarizing output, plotting, goodness-of-fit and convergence diagnostics.

- GeNIe & SMILE by BayesFusion — SMILE is a library for BNs, DBNs, IDs, equation-based models, and learning. SMILE is available for multiple platforms, is written in C++, and has wrappers for several programming environments through Java and .NET interfaces. GeNIe is graphical front end for SMILE, running natively on Windows and on Linux and Mac under Wine.

- SamIam (website), a Java-based system with GUI and Java API

- Bayes Server - User Interface and API for Bayesian networks, includes support for time series and sequences

- Blip - Blip is a web interface that offers structural learning of Bayesian networks directly from discrete data. It can handle datasets with thousands of variables, and offers both unbounded and treewidth-bounded structure learning.

- Belief and Decision Networks on AIspace

- BayesiaLab by Bayesia

- Hugin

- AgenaRisk

- Netica by Norsys

- dVelox by Apara Software

- System Modeler by Inatas AB

- UnBBayes by GIA-UnB (Intelligence Artificial Group - University of Brasilia)

- Face2gene using the Facial Dysmorphology Novel Analysis (FDNA) technology

- Uninet — Continuous Bayesian networks modelling continuous variables, with a wide range of parametric and non-parametric marginal distributions, and dependence with copula. Hybrid discrete continuous models are also supported. Free for non-commercial use. Developed by LightTwist Software.

- Tetrad, an open-source project written in Java and developed by the Department of Philosophy at Carnegie Mellon University, that deals with causal models and statistical data.

- Dezide

Semantics of Bayesian Networks

The simplest conditional independence relationship encoded in a Bayesian network can be stated as follows: a node is independent of its ancestors given its parents, where the ancestor/parent relationship is with respect to some fixed topological ordering of the nodes.

In the sprinkler example above, by the chain rule of probability, the joint probability of all the nodes in the graph above is

P(C, S, R, W) = P(C) * P(S|C) * P(R|C,S) * P(W|C,S,R)

By using conditional independence relationships, we can rewrite this as

P(C, S, R, W) = P(C) * P(S|C) * P(R|C) * P(W|S,R)

where we were allowed to simplify the third term because R is independent of S given its parent C, and the last term because W is independent of C given its parents S and R. We can see that the conditional independence relationships allow us to represent the joint more compactly. Here the savings are minimal, but in general, if we had n binary nodes, the full joint would require $O(2^n)$ space to represent, but the factored form would require $O(n\,2^k)$ space to represent, where k is the maximum fan-in of a node. And fewer parameters makes learning easier.

The intuitive meaning of an arrow from a parent to a child is that the parent directly influences the child. The direction of this influence is often taken to represent casual influence. The conditional probabilities give the strength of causal influence. A 0 or 1 in a CPT represents a deterministic influence.

$$P(E = t|T = t, C = t) = 1 \qquad P(E = t|T = t, L = f) = 1$$
$$P(E = t|T = f, L = t) = 1 \qquad P(E = t|T = f, L = f) = 0$$

Ecomposing Joint Distributions

A joint distribution can always be broken down into a product of conditional probabilities using repeated applications of the product rule.

$$P(A,T,E,L,S,B,D,X) = P(X|A,T,E,L,S,B,D)P(D|A,T,E,L,S,B)$$
$$P(B|A,T,E,L,S)P(S|A,T,E,L)P(L|A,T,E)P(E|A,T)P(T|A)P(A)$$

We can order the variables however we like:

$$P(A,T,E,L,S,B,D,X) = P(X|A,T,E,L,S,B,D)P(D|A,T,E,L,S,B)$$
$$P(E|A,T,L,S,B)P(B|A,T,L,S)P(L|A,T,S)P(S|A,T)P(T|A)P(A)$$

Conditional Independence in Bayes Net

A Bayes net represents the assumption that each node is conditionally independent of all its non-descendants given its parents.

So for example,

$$P(E|A,T,L,S,B) = P(E|T,L)$$

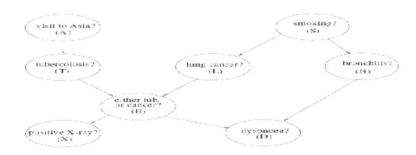

Note that, a node is NOT independent of its descendants given its parents. Generally,

$$P(E|A,T,L,S,B,X) \neq P(E|T,L)$$

Variable Ordering in Bayes Net

The conditional independence assumptions expressed by a Bayes net allow a compact representation of the joint distribution. First note that the Bayes net imposes a partial order on nodes: X <= Y iff X is a descendant of Y. We can always break down the joint so that the conditional probability factor for a node only has non-descendants in the condition.

$$P(A,T,E,L,S,B,D,X) = P(X|A,T,E,L,S,B,D)P(D|A,T,E,L,S,B)$$
$$P(E|A,T,L,S,B)P(B|A,T,L,S)P(L|A,T,S)P(S|A,T)P(T|A)P(A)$$

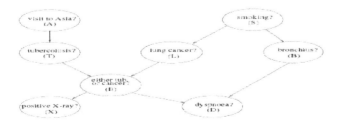

The Joint Distribution as a Product of CPTs

Because each node is conditionally independent of all its nondescendants given its parents, and because we can write the joint appropriately we have:

$$P(A,T,E,L,S,B,D,X) =$$
$$P(X|A,T,E,L,S,B,D)P(D|A,T,E,L,S,B)$$
$$P(E|A,T,L,S,B)P(B|A,T,L,S)$$
$$P(L|A,T,S)P(S|A,T)P(T|A)P(A)$$
$$=$$
$$P(X|E)P(D|E,B)P(E|T,L)P(B|S)$$
$$P(L|S)P(S)P(T|A)P(A)$$

So the CPTs Determine the Full Joint Distribution.

In short,

$$P(X_1, X_2, \ldots, X_n) = \prod_{i=1}^{n} P(X_i|Parents(X_i))$$

Bayesian Networks allow a compact representation of the probability distributions. An unstructured table representation of the "medical expert system" joint would require

$2^8 - 1 = 255$ numbers. With the structure imposed by the conditional independence assumptions this reduces to 18 numbers. Structure also allows efficient inference — of which more later.

Conditional Independence and d-separation in a Bayesian Network

We can have conditional independence relations between sets of random variables. In the Medical Expert System Bayesian net, {X, D} is independent of {A, T, L, S} given {E,B} which means:

$$P(X,D|\ E,B) = P(X,D|E,B,A,T,L,S)$$

equivalently

$$P(X,\ D,\ A,T,L,\ S|\ E,B) = P(A,T,L,S|E,B)P(X,D|\ E,B)$$

We need a way of checking for these conditional independence relations

Conditional independence can be checked uing the d-separation property of the Bayes net directed acyclic graph. d-separation is short for direction-dependent separation.

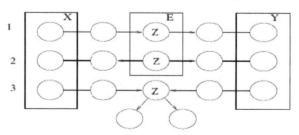

If E d-separates X and Y then X and Y are conditionally independent given E.

E d-separates X and Y if every *undirected path* from a node in X to a node in Y is blocked given E.

Defining d-separation

A path is blocked given a set of nodes E if there is a node Z on the path for which one of these three conditions holds:

1. Z is in E and Z has one arrow on the path coming in and one arrow going out.

2. Z is in E and Z has both path arrows leading out.

3. Neither Z nor any descendant of Z is in E, and both path arrows lead in to Z.

Building a Bayes Net: The Family Out? Example

We start with a natural language description of the situation to be modeled:

I want to know if my family is at home as I approach the house. Often my wife leaves on a light when she goes out, but also sometimes if she is expecting a guest. When nobody is home the dog is put in the back yard, but he is also put there when he has bowel trouble. If the dog is in the back yard, I will hear her barking, but I may be confused by other dogs barking.

Building the Bayes net involves the following steps.

We build Bayes nets to get probabilities concerning what we don't know given what we do know. What we don't know is not observable. These are called hypothesis events – we need to know what are the hypothesis events in a problem?

Recall that a Bayesian network is composed of related (random) variables, and that a variable incorporates an exhaustive set of mutually exclusive events - one of its events is true. How shall we represent the two hypothesis events in a problem?

Variables whose values are observable and which are relevant to the hypothesis events are called information variables. What are the information variables in a problem?

In this problem we have three variables, what is the causal structure between them? Actually, the whole notion of 'cause' let alone 'determining causal structure' is very controversial. Often (but not always) your intuitive notion of causality will help you.

Sometimes we need mediating variables which are neither information variables or hypothesis variables to represent causal structures.

Learning of Bayesian Network Parameters

One needs to specify two things to describe a BN: the graph topology (structure) and the parameters of each CPT. It is possible to learn both of these from data. However, learning structure is much harder than learning parameters. Also, learning when some of the nodes are hidden, or we have missing data, is much harder than when everything is observed. This gives rise to 4 cases:

Structure	Observability	Method
Known	Full	Maximum Likelihood Estimation
Known	Partial	EM (or gradient ascent)
Unknown	Full	Search through model space
Unknown	Partial	EM + search through model space

We discuss below only the first case.

Known structure, full observability-

We assume that the goal of learning in this case is to find the values of the parameters of each CPT which maximizes the likelihood of the training data, which contains N cases (assumed to be independent). The normalized log-likelihood of the training set D is a sum of terms, one for each node:

$$L = \frac{1}{N} \sum_{i=1}^{m} \sum_{i=1}^{S} \log P(X_i | Pa(X_i)_1 D_l).$$

We see that the log-likelihood scoring function decomposes according to the structure of the graph, and hence we can maximize the contribution to the log-likelihood of each node independently (assuming the parameters in each node are independent of the other nodes). In cases where N is small compared to the number of parameters that require fitting, we can use a numerical prior to regularize the problem. In this case, we call the estimates Maximum A Posterori (MAP) estimates, as opposed to Maximum Likelihood (ML) estimates.

Consider estimating the Conditional Probability Table for the W node. If we have a set of training data, we can just count the number of times the grass is wet when it is raining and the sprinler is on, N(W=1,S=1,R=1), the number of times the grass is wet when it is raining and the sprinkler is off, N(W=1,S=0,R=1), etc. Given these counts (which are the sufficient statistics), we can find the Maximum Likelihood Estimate of the CPT as follows:

$$\Pr(W = \omega | S = s_1 R = r) \approx N(W = \omega_1 S = s_1 R = r) / N(S = s_1 R = r)$$

where the denominator is N(S=s,R=r) = N(W=0,S=s,R=r) + N(W=1,S=s,R=r). Thus "learning" just amounts to counting (in the case of multinomial distributions). For Gaussian nodes, we can compute the sample mean and variance, and use linear regression to estimate the weight matrix. For other kinds of distributions, more complex procedures are necessary.

As is well known from the HMM literature, ML estimates of CPTs are prone to sparse data problems, which can be solved by using (mixtures of) Dirichlet priors (pseudo counts). This results in a Maximum A Posteriori (MAP) estimate. For Gaussians, we can use a Wishart prior, etc.

Bayesian Inference

Bayesian inference is a method of statistical inference in which Bayes' theorem is used to update the probability for a hypothesis as more evidence or information becomes available. Bayesian inference is an important technique in statistics, and especially in mathematical statistics. Bayesian updating is particularly important in the dynamic analysis of a sequence of data. Bayesian inference has found application in a wide range of activities, including science, engineering, philosophy, medicine, sport, and law. In the philosophy of decision theory, Bayesian inference is closely related to subjective probability, often called "Bayesian probability".

Introduction to Bayes' Rule

In the table below, the values w, x, y and z give the relative weights of each corresponding condition and case. The figures denote the cells of the table involved in each metric, the probability being the fraction of each figure that is shaded. This shows that P(A|B) P(B) = P(B|A) P(A) i.e. P(A|B) = P(B|A) P(A)/P(B) . Similar reasoning can be used to show that P(Ā|B) = P(B|Ā) P(Ā)/P(B) etc.

Relative size	Case B	Case B̄	Total
Condition A	w	x	$w+x$
Condition Ā	y	z	$y+z$
Total	$w+y$	$x+z$	$w+x+y+z$

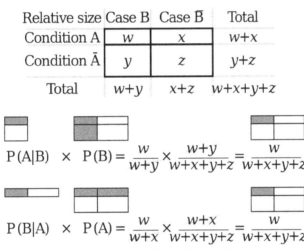

$$P(A|B) \times P(B) = \frac{w}{w+y} \times \frac{w+y}{w+x+y+z} = \frac{w}{w+x+y+z}$$

$$P(B|A) \times P(A) = \frac{w}{w+x} \times \frac{w+x}{w+x+y+z} = \frac{w}{w+x+y+z}$$

A geometric visualisation of Bayes' theorem

Formal

Bayesian inference derives the posterior probability as a consequence of two antecedents, a prior probability and a "likelihood function" derived from a statistical model for the observed data. Bayesian inference computes the posterior probability according to Bayes' theorem:

$$P(H|E) = \frac{P(E|H) \cdot P(H)}{P(E)}$$

where

- | means "event conditional on" (so that $(A|B)$ means *A given B*).

- *H* stands for any *hypothesis* whose probability may be affected by data (called *evidence* below). Often there are competing hypotheses, and the task is to determine which is the most probable.

- the *evidence* *E* corresponds to new data that were not used in computing the prior probability.

- $P(H)$, the *prior probability*, is the estimate of the probability of the hypothesis H *before* the data E, the current evidence, is observed.

- $P(H\,|\,E)$, the *posterior probability*, is the probability of H *given* E, i.e., *after* E is observed. This is what we want to know: the probability of a hypothesis *given* the observed evidence.

- $P(E\,|\,H)$ is the probability of observing E *given* H. As a function of E with H fixed, this is the *likelihood* – it indicates the compatibility of the evidence with the given hypothesis. The likelihood function is a function of the evidence, E, while the posterior probability is a function of the hypothesis, H.

- $P(E)$ is sometimes termed the marginal likelihood or "model evidence". This factor is the same for all possible hypotheses being considered (as is evident from the fact that the hypothesis H H does not appear anywhere in the symbol, unlike for all the other factors), so this factor does not enter into determining the relative probabilities of different hypotheses.

For different values of H, only the factors $P(H)$ and $P(E\,|\,H)$, both in the numerator, affect the value of $P(H\,|\,E)$ – the posterior probability of a hypothesis is proportional to its prior probability (its inherent likeliness) and the newly acquired likelihood (its compatibility with the new observed evidence).

Bayes' rule can also be written as follows:

$$P(H\,|\,E) = \frac{P(E\,|\,H)}{P(E)} \cdot P(H)$$

where the factor $\dfrac{P(E\,|\,H)}{P(E)}$ can be interpreted as the impact of E on the probability of H.

Informal

If the evidence does not match up with a hypothesis, one should reject the hypothesis. But if a hypothesis is extremely unlikely *a priori*, one should also reject it, even if the evidence does appear to match up. For example, if one does not know whether the newborn baby next door is a boy or a girl, the color of decorations on the crib in front of the door may support the hypothesis of one gender or the other; but if in front of that door, instead of the crib, a dog kennel is found, the posterior probability that the family next door gave birth to a dog remains small in spite of the "evidence", since one's prior belief in such a hypothesis was already extremely small.

The critical point about Bayesian inference, then, is that it provides a principled way of combining new evidence with prior beliefs, through the application of Bayes' rule. (Contrast this with frequentist inference, which relies only on the evidence as a whole, with no reference to prior beliefs.)

Furthermore, Bayes' rule can be applied iteratively: after observing some evidence, the resulting posterior probability can then be treated as a prior probability, and a new posterior probability computed from new evidence. This allows for Bayesian principles to be applied to various kinds of evidence, whether viewed all at once or over time. This procedure is termed "Bayesian updating".

Alternatives to Bayesian Updating

Bayesian updating is widely used and computationally convenient. However, it is not the only updating rule that might be considered rational.

Ian Hacking noted that traditional "Dutch book" arguments did not specify Bayesian updating: they left open the possibility that non-Bayesian updating rules could avoid Dutch books. Hacking wrote "And neither the Dutch book argument, nor any other in the personalist arsenal of proofs of the probability axioms, entails the dynamic assumption. Not one entails Bayesianism. So the personalist requires the dynamic assumption to be Bayesian. It is true that in consistency a personalist could abandon the Bayesian model of learning from experience. Salt could lose its savour."

Indeed, there are non-Bayesian updating rules that also avoid Dutch books (as discussed in the literature on "probability kinematics") following the publication of Richard C. Jeffrey's rule, which applies Bayes' rule to the case where the evidence itself is assigned a probability. The additional hypotheses needed to uniquely require Bayesian updating have been deemed to be substantial, complicated, and unsatisfactory.

Formal Description of Bayesian Inference

Definitions

- x, a data point in general. This may in fact be a vector of values.

- θ, the parameter of the data point's distribution, i.e., $x \sim p(x \mid \theta)$. This may in fact be a vector of parameters.

- α, the hyperparameter of the parameter, i.e., $\theta \sim p(\theta \mid \alpha)$. This may in fact be a vector of hyperparameters.

- $\mathbf{X}$, a set of n observed data points, i.e., $x_1, \ldots, x_n$.

- $\tilde{x}$, a new data point whose distribution is to be predicted.

Bayesian Inference

- The prior distribution is the distribution of the parameter(s) before any data is observed, i.e. $p(\theta \mid \alpha)$.

- The prior distribution might not be easily determined. In this case, we can use the Jeffreys prior to obtain the posterior distribution before updating them with newer observations.

- The sampling distribution is the distribution of the observed data conditional on its parameters, i.e. $p(\mathbf{X} \mid \theta)$. This is also termed the likelihood, especially when viewed as a function of the parameter(s), sometimes written $L(\theta \mid \mathbf{X}) = p(\mathbf{X} \mid \theta)$.

- The marginal likelihood (sometimes also termed the *evidence*) is the distribution of the observed data marginalized over the parameter(s), i.e. $p(\mathbf{X} \mid \alpha) = \int_{\theta} p(\mathbf{X} \mid \theta) p(\theta \mid \alpha) \mathrm{d}\theta$.

- The posterior distribution is the distribution of the parameter(s) after taking into account the observed data. This is determined by Bayes' rule, which forms the heart of Bayesian inference:

$$p(\theta \mid \mathbf{X}, \alpha) = \frac{p(\mathbf{X} \mid \theta)p(\theta \mid \alpha)}{p(\mathbf{X} \mid \alpha)} \propto p(\mathbf{X} \mid \theta)p(\theta \mid \alpha)$$

Note that this is expressed in words as "posterior is proportional to likelihood times prior", or sometimes as "posterior = likelihood times prior, over evidence".

Bayesian Prediction

- The posterior predictive distribution is the distribution of a new data point, marginalized over the posterior:

$$p(\tilde{x} \mid \mathbf{X}, \alpha) = \int_{\theta} p(\tilde{x} \mid \theta)p(\theta \mid \mathbf{X}, \alpha)\,\mathrm{d}\theta$$

- The prior predictive distribution is the distribution of a new data point, marginalized over the prior:

$$p(\tilde{x} \mid \alpha) = \int_{\theta} p(\tilde{x} \mid \theta)p(\theta \mid \alpha)\,\mathrm{d}\theta$$

Bayesian theory calls for the use of the posterior predictive distribution to do predictive inference, i.e., to predict the distribution of a new, unobserved data point. That is, instead of a fixed point as a prediction, a distribution over possible points is returned. Only this way is the entire posterior distribution of the parameter(s) used. By comparison, prediction in frequentist statistics often involves finding an optimum point estimate of the parameter(s)—e.g., by maximum likelihood or maximum a posteriori estimation (MAP)—and then plugging this estimate into the formula for the distribution of a data point. This has the disadvantage that it does not account for any uncertainty in the value of the parameter, and hence will underestimate the variance of the predictive distribution.

(In some instances, frequentist statistics can work around this problem. For example, confidence intervals and prediction intervals in frequentist statistics when constructed from a normal distribution with unknown mean and variance are constructed using a Student's t-distribution. This correctly estimates the variance, due to the fact that (1) the average of normally distributed random variables is also normally distributed; (2) the predictive distribution of a normally distributed data point with unknown mean and variance, using conjugate or uninformative priors, has a student's t-distribution. In Bayesian statistics, however, the posterior predictive distribution can always be determined exactly—or at least, to an arbitrary level of precision, when numerical methods are used.)

Note that both types of predictive distributions have the form of a compound probability distribution (as does the marginal likelihood). In fact, if the prior distribution is a conjugate prior, and hence the prior and posterior distributions come from the same family, it can easily be seen that both prior and posterior predictive distributions also come from the same family of compound distributions. The only difference is that the posterior predictive distribution uses the updated values of the hyperparameters (applying the Bayesian update rules given in the conjugate prior article), while the prior predictive distribution uses the values of the hyperparameters that appear in the prior distribution.

Inference Over Exclusive and Exhaustive Possibilities

If evidence is simultaneously used to update belief over a set of exclusive and exhaustive propositions, Bayesian inference may be thought of as acting on this belief distribution as a whole.

General Formulation

Suppose a process is generating independent and identically distributed events E_n, but the probability distribution is unknown. Let the event space $\grave{U}$ represent the current state of belief for this process. Each model is represented by event M_m. The conditional probabilities $P(E_n \mid M_m)$ are specified to define the models. $P(M_m)$ is the degree of belief in M_m. Before the first inference step, $\{P(M_m)\}$ is a set of *initial prior probabilities*. These must sum to 1, but are otherwise arbitrary.

Diagram illustrating event space $\grave{U}$ in general formulation of Bayesian inference. Although this diagram shows discrete models and events, the continuous case may be visualized similarly using probability densities.

Suppose that the process is observed to generate $E \in \{E_n\}$. For each $M \in \{M_m\}$, the prior $P(M)$ is updated to the posterior $P(M \mid E)$. From Bayes' theorem:

$$P(M \mid E) = \frac{P(E \mid M)}{\sum_m P(E \mid M_m)P(M_m)} \cdot P(M)$$

Upon observation of further evidence, this procedure may be repeated.

Multiple Observations

For a sequence of independent and identically distributed observations $\mathbf{E} = (e_1, \ldots, e_n)$, it can be shown by induction that repeated application of the above is equivalent to

$$P(M \mid \mathbf{E}) = \frac{P(\mathbf{E} \mid M)}{\sum_m P(\mathbf{E} \mid M_m)P(M_m)} \cdot P(M)$$

Where

$$P(\mathbf{E} \mid M) = \prod_k P(e_k \mid M).$$

Parametric Formulation

By parameterizing the space of models, the belief in all models may be updated in a single step. The distribution of belief over the model space may then be thought of as a distribution of belief over the parameter space. The distributions in this section are expressed as continuous, represented by probability densities, as this is the usual situation. The technique is however equally applicable to discrete distributions.

Let the vector θ span the parameter space. Let the initial prior distribution over θ be $p(\theta \mid \alpha)$, where α is a set of parameters to the prior itself, or *hyperparameters*. Let $\mathbf{E} = (e_1, \ldots, e_n)$ be a sequence of independent and identically distributed event observations, where all e_i are distributed as $p(e \mid \theta)$ for some θ. Bayes' theorem is applied to find the posterior distribution over θ:

$$p(\theta \mid \mathbf{E}, \alpha) = \frac{p(\mathbf{E} \mid \theta, \alpha)}{p(\mathbf{E} \mid \alpha)} \cdot p(\theta \mid \alpha) = \frac{p(\mathbf{E} \mid \theta, \alpha)}{\int_{\theta} p(\mathbf{E} \mid \theta, \alpha) p(\theta \mid \alpha) d\theta} \cdot p(\theta \mid \alpha)$$

Where

$$p(\mathbf{E} \mid \theta, \alpha) = \prod_{k} p(e_k \mid \theta)$$

Mathematical Properties

Interpretation of Factor

$\frac{P(E \mid M)}{P(E)} > 1 \Rightarrow P(E \mid M) > P(E)$. That is, if the model were true, the evidence would be more likely than is predicted by the current state of belief. The reverse applies for a decrease in belief. If the belief does not change, $\frac{P(E \mid M)}{P(E)} = 1 \Rightarrow P(E \mid M) = P(E)$. That is, the evidence is independent of the model. If the model were true, the evidence would be exactly as likely as predicted by the current state of belief.

Cromwell's Rule

If $P(M) = 0$ then $P(M \mid E) = 0$. If $P(M) = 1$, then $P(M \mid E) = 1$. This can be interpreted to mean that hard convictions are insensitive to counter-evidence.

The former follows directly from Bayes' theorem. The latter can be derived by applying the first rule to the event "not M" in place of "M", yielding "if $1 - P(M) = 0$, then $1 - P(M \mid E) = 0$", from which the result immediately follows.

Asymptotic Behaviour of Posterior

Consider the behaviour of a belief distribution as it is updated a large number of times with independent and identically distributed trials. For sufficiently nice prior probabilities, the Bernstein-von Mises theorem gives that in the limit of infinite trials, the posterior converges to a

Gaussian distribution independent of the initial prior under some conditions firstly outlined and rigorously proven by Joseph L. Doob in 1948, namely if the random variable in consideration has a finite probability space. The more general results were obtained later by the statistician David A. Freedman who published in two seminal research papers in 1963 and 1965 when and under what circumstances the asymptotic behaviour of posterior is guaranteed. His 1963 paper treats, like Doob (1949), the finite case and comes to a satisfactory conclusion. However, if the random variable has an infinite but countable probability space (i.e., corresponding to a die with infinite many faces) the 1965 paper demonstrates that for a dense subset of priors the Bernstein-von Mises theorem is not applicable. In this case there is almost surely no asymptotic convergence. Later in the 1980s and 1990s Freedman and Persi Diaconis continued to work on the case of infinite countable probability spaces. To summarise, there may be insufficient trials to suppress the effects of the initial choice, and especially for large (but finite) systems the convergence might be very slow.

Conjugate Priors

In parameterized form, the prior distribution is often assumed to come from a family of distributions called conjugate priors. The usefulness of a conjugate prior is that the corresponding posterior distribution will be in the same family, and the calculation may be expressed in closed form.

Estimates of Parameters and Predictions

It is often desired to use a posterior distribution to estimate a parameter or variable. Several methods of Bayesian estimation select measurements of central tendency from the posterior distribution.

For one-dimensional problems, a unique median exists for practical continuous problems. The posterior median is attractive as a robust estimator.

If there exists a finite mean for the posterior distribution, then the posterior mean is a method of estimation.

$$\tilde{\theta} = \mathrm{E}[\theta] = \int_{\theta} \theta\, p(\theta \mid \mathbf{X}, \alpha) d\theta$$

Taking a value with the greatest probability defines maximum *a posteriori* (MAP) estimates:

$$\{\theta_{\mathrm{MAP}}\} \subset \arg \max_{\theta} p(\theta \mid \mathbf{X}, \alpha).$$

There are examples where no maximum is attained, in which case the set of MAP estimates is empty.

There are other methods of estimation that minimize the posterior *risk* (expected-posterior loss) with respect to a loss function, and these are of interest to statistical decision theory using the sampling distribution ("frequentist statistics").

The posterior predictive distribution of a new observation $\tilde{x}$ (that is independent of previous observations) is determined by

$$p(\tilde{x} \mid \mathbf{X}, \alpha) = \int_{\theta} p(\tilde{x}, \theta \mid \mathbf{X}, \alpha) d\theta = \int_{\theta} p(\tilde{x} \mid \theta) p(\theta \mid \mathbf{X}, \alpha) d\theta.$$

Examples

Probability of a Hypothesis

Suppose there are two full bowls of cookies. Bowl #1 has 10 chocolate chip and 30 plain cookies, while bowl #2 has 20 of each. Our friend Fred picks a bowl at random, and then picks a cookie at random. We may assume there is no reason to believe Fred treats one bowl differently from another, likewise for the cookies. The cookie turns out to be a plain one. How probable is it that Fred picked it out of bowl #1?

Intuitively, it seems clear that the answer should be more than a half, since there are more plain cookies in bowl #1. The precise answer is given by Bayes' theorem. Let H_1 correspond to bowl #1, and H_2 to bowl #2. It is given that the bowls are identical from Fred's point of view, thus $P(H_1) = P(H_2)$, and the two must add up to 1, so both are equal to 0.5. The event E is the observation of a plain cookie. From the contents of the bowls, we know that $P(E \mid H_1) = 30/40 = 0.75$ and $P(E \mid H_2) = 20/40 = 0.5$. Bayes' formula then yields

$$P(H_1 \mid E) = \frac{P(E \mid H_1)P(H_1)}{P(E \mid H_1)P(H_1) + P(E \mid H_2)P(H_2)} = \frac{0.75 \times 0.5}{0.75 \times 0.5 + 0.5 \times 0.5} = 0.6$$

Before we observed the cookie, the probability we assigned for Fred having chosen bowl #1 was the prior probability, $P(H_1)$, which was 0.5. After observing the cookie, we must revise the probability to $P(H_1 \mid E)$, which is 0.6.

Making a Prediction

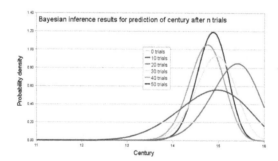

Example results for archaeology example. This simulation was generated using c=15.2.

An archaeologist is working at a site thought to be from the medieval period, between the 11th century to the 16th century. However, it is uncertain exactly when in this period the site was inhabited. Fragments of pottery are found, some of which are glazed and some of which are decorated. It is expected that if the site were inhabited during the early medieval period, then 1% of the pottery would be glazed and 50% of its area decorated, whereas if it had been inhabited in the late medieval period then 81% would be glazed and 5% of its area decorated. How confident can the archaeologist be in the date of inhabitation as fragments are unearthed?

The degree of belief in the continuous variable C (century) is to be calculated, with the discrete set of events $\{GD, G\bar{D}, \bar{G}D, \bar{G}\bar{D}\}$ as evidence. Assuming linear variation of glaze and decoration with time, and that these variables are independent,

$$P(E = GD \mid C = c) = (0.01 + \frac{0.81 - 0.01}{16 - 11}(c - 11))(0.5 - \frac{0.5 - 0.05}{16 - 11}(c - 11))$$

$$P(E = G\bar{D} \mid C = c) = (0.01 + \frac{0.81 - 0.01}{16 - 11}(c - 11))(0.5 + \frac{0.5 - 0.05}{16 - 11}(c - 11))$$

$$P(E = \bar{G}D \mid C = c) = ((1 - 0.01) - \frac{0.81 - 0.01}{16 - 11}(c - 11))(0.5 - \frac{0.5 - 0.05}{16 - 11}(c - 11))$$

$$P(E = \bar{G}\bar{D} \mid C = c) = ((1 - 0.01) - \frac{0.81 - 0.01}{16 - 11}(c - 11))(0.5 + \frac{0.5 - 0.05}{16 - 11}(c - 11))$$

Assume a uniform prior of $f_C(c) = 0.2$, and that trials are independent and identically distributed. When a new fragment of type e is discovered, Bayes' theorem is applied to update the degree of belief for each c:

$$f_C(c \mid E = e) = \frac{P(E = e \mid C = c)}{P(E = e)} f_C(c) = \frac{P(E = e \mid C = c)}{\int_{11}^{16} P(E = e \mid C = c) f_C(c) dc} f_C(c)$$

A computer simulation of the changing belief as 50 fragments are unearthed is shown on the graph. In the simulation, the site was inhabited around 1420, or $c = 15.2$. By calculating the area under the relevant portion of the graph for 50 trials, the archaeologist can say that there is practically no chance the site was inhabited in the 11th and 12th centuries, about 1% chance that it was inhabited during the 13th century, 63% chance during the 14th century and 36% during the 15th century. Note that the Bernstein-von Mises theorem asserts here the asymptotic convergence to the "true" distribution because the probability space corresponding to the discrete set of events $\{GD, G\bar{D}, \bar{G}D, \bar{G}\bar{D}\}$ is finite.

In Frequentist Statistics and Decision Theory

A decision-theoretic justification of the use of Bayesian inference was given by Abraham Wald, who proved that every unique Bayesian procedure is admissible. Conversely, every admissible statistical procedure is either a Bayesian procedure or a limit of Bayesian procedures.

Wald characterized admissible procedures as Bayesian procedures (and limits of Bayesian procedures), making the Bayesian formalism a central technique in such areas of frequentist inference as parameter estimation, hypothesis testing, and computing confidence intervals. For example:

- "Under some conditions, all admissible procedures are either Bayes procedures or limits of Bayes procedures (in various senses). These remarkable results, at least in their original form, are due essentially to Wald. They are useful because the property of being Bayes is easier to analyze than admissibility."

- "In decision theory, a quite general method for proving admissibility consists in exhibiting a procedure as a unique Bayes solution."

- "In the first chapters of this work, prior distributions with finite support and the corresponding Bayes procedures were used to establish some of the main theorems relating to the comparison of experiments. Bayes procedures with respect to more general prior distributions have played a very important role in the development of statistics, including its asymptotic theory." "There are many problems where a glance at posterior distributions, for suitable priors, yields immediately interesting information. Also, this technique can hardly be avoided in sequential analysis."

- "A useful fact is that any Bayes decision rule obtained by taking a proper prior over the whole parameter space must be admissible"

- "An important area of investigation in the development of admissibility ideas has been that of conventional sampling-theory procedures, and many interesting results have been obtained."

Model Selection

Computer Applications

Bayesian inference has applications in artificial intelligence and expert systems. Bayesian inference techniques have been a fundamental part of computerized pattern recognition techniques since the late 1950s. There is also an ever growing connection between Bayesian methods and simulation-based Monte Carlo techniques since complex models cannot be processed in closed form by a Bayesian analysis, while a graphical model structure *may* allow for efficient simulation algorithms like the Gibbs sampling and other Metropolis–Hastings algorithm schemes. Recently Bayesian inference has gained popularity amongst the phylogenetics community for these reasons; a number of applications allow many demographic and evolutionary parameters to be estimated simultaneously.

As applied to statistical classification, Bayesian inference has been used in recent years to develop algorithms for identifying e-mail spam. Applications which make use of Bayesian inference for spam filtering include CRM114, DSPAM, Bogofilter, SpamAssassin, SpamBayes, Mozilla, XE-AMS, and others. Spam classification is treated in more detail in the article on the naive Bayes classifier.

Solomonoff's Inductive inference is the theory of prediction based on observations; for example, predicting the next symbol based upon a given series of symbols. The only assumption is that the environment follows some unknown but computable probability distribution. It is a formal inductive framework that combines two well-studied principles of inductive inference: Bayesian statistics and Occam's Razor. Solomonoff's universal prior probability of any prefix p of a computable sequence x is the sum of the probabilities of all programs (for a universal computer) that compute something starting with p. Given some p and any computable but unknown probability distribution from which x is sampled, the universal prior and Bayes' theorem can be used to predict the yet unseen parts of x in optimal fashion.

In the Courtroom

Bayesian inference can be used by jurors to coherently accumulate the evidence for and against a defendant, and to see whether, in totality, it meets their personal threshold for 'beyond a reasonable doubt'. Bayes' theorem is applied successively to all evidence presented, with the posterior from one stage becoming the prior for the next. The benefit of a Bayesian approach is that it gives the juror an unbiased, rational mechanism for combining evidence. It may be appropriate to explain Bayes' theorem to jurors in odds form, as betting odds are more widely understood than probabilities. Alternatively, a logarithmic approach, replacing multiplication with addition, might be easier for a jury to handle.

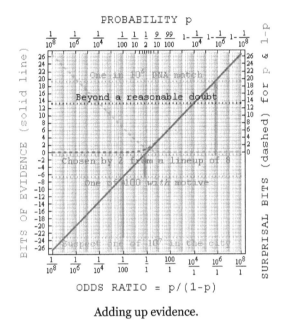

Adding up evidence.

If the existence of the crime is not in doubt, only the identity of the culprit, it has been suggested that the prior should be uniform over the qualifying population. For example, if 1,000 people could have committed the crime, the prior probability of guilt would be 1/1000.

The use of Bayes' theorem by jurors is controversial. In the United Kingdom, a defence expert witness explained Bayes' theorem to the jury in *R v Adams*. The jury convicted, but the case went to appeal on the basis that no means of accumulating evidence had been provided for jurors who did not wish to use Bayes' theorem. The Court of Appeal upheld the conviction, but it also gave the opinion that "To introduce Bayes' Theorem, or any similar method, into a criminal trial plunges the jury into inappropriate and unnecessary realms of theory and complexity, deflecting them from their proper task."

Gardner-Medwin argues that the criterion on which a verdict in a criminal trial should be based is *not* the probability of guilt, but rather the *probability of the evidence, given that the defendant is innocent* (akin to a frequentist p-value). He argues that if the posterior probability of guilt is to be computed by Bayes' theorem, the prior probability of guilt must be known. This will depend on the incidence of the crime, which is an unusual piece of evidence to consider in a criminal trial. Consider the following three propositions:

A The known facts and testimony could have arisen if the defendant is guilty

B The known facts and testimony could have arisen if the defendant is innocent

C The defendant is guilty.

Gardner-Medwin argues that the jury should believe both A and not-B in order to convict. A and not-B implies the truth of C, but the reverse is not true. It is possible that B and C are both true, but in this case he argues that a jury should acquit, even though they know that they will be letting some guilty people go free.

Bayesian Epistemology

Bayesian epistemology is a movement that advocates for Bayesian inference as a means of justifying the rules of inductive logic.

Karl Popper and David Miller have rejected the alleged rationality of Bayesianism, i.e. using Bayes rule to make epistemological inferences: It is prone to the same vicious circle as any other justificationist epistemology, because it presupposes what it attempts to justify. According to this view, a rational interpretation of Bayesian inference would see it merely as a probabilistic version of falsification, rejecting the belief, commonly held by Bayesians, that high likelihood achieved by a series of Bayesian updates would prove the hypothesis beyond any reasonable doubt, or even with likelihood greater than 0.

Other

- The scientific method is sometimes interpreted as an application of Bayesian inference. In this view, Bayes' rule guides (or should guide) the updating of probabilities about hypotheses conditional on new observations or experiments.

- Bayesian search theory is used to search for lost objects.

- Bayesian inference in phylogeny

- Bayesian tool for methylation analysis

- Bayesian approaches to brain function investigate the brain as a Bayesian mechanism.

- Bayesian inference in ecological studies

Bayes and Bayesian Inference

The problem considered by Bayes in Proposition 9 of his essay, "An Essay towards solving a Problem in the Doctrine of Chances", is the posterior distribution for the parameter a (the success rate) of the binomial distribution.

History

The term *Bayesian* refers to Thomas Bayes (1702–1761), who proved a special case of what is now called Bayes' theorem. However, it was Pierre-Simon Laplace (1749–1827) who introduced a general version of the theorem and used it to approach problems in celestial mechanics, medical statistics, reliability, and jurisprudence. Early Bayesian inference, which used uniform priors following Laplace's principle of insufficient reason, was called "inverse probability" (because it infers backwards from observations to parameters, or from effects to causes). After the 1920s, "inverse probability" was largely supplanted by a collection of methods that came to be called frequentist statistics.

In the 20th century, the ideas of Laplace were further developed in two different directions, giving rise to *objective* and *subjective* currents in Bayesian practice. In the objective or "non-informative" current, the statistical analysis depends on only the model assumed, the data analyzed, and the method assigning the prior, which differs from one objective Bayesian to another objective Bayesian. In the subjective or "informative" current, the specification of the prior depends on the belief (that is, propositions on which the analysis is prepared to act), which can summarize information from experts, previous studies, etc.

In the 1980s, there was a dramatic growth in research and applications of Bayesian methods, mostly attributed to the discovery of Markov chain Monte Carlo methods, which removed many of the computational problems, and an increasing interest in nonstandard, complex applications. Despite growth of Bayesian research, most undergraduate teaching is still based on frequentist statistics. Nonetheless, Bayesian methods are widely accepted and used, such as for example in the field of machine learning.

References

- Pearl, Judea; Russell, Stuart (November 2002). "Bayesian Networks". In Arbib, Michael A. Handbook of Brain Theory and Neural Networks. Cambridge, Massachusetts: Bradford Books (MIT Press). pp. 157–160. ISBN 0-262-01197-2

- Jøsang, A., 2001, "A logic for uncertain probabilities," International Journal of Uncertainty, Fuzziness and Knowledge-Based Systems 9(3):279-311

- Castillo, Enrique; Gutiérrez, José Manuel; Hadi, Ali S. (1997). "Learning Bayesian Networks". Expert Systems and Probabilistic Network Models. Monographs in computer science. New York: Springer-Verlag. pp. 481–528. ISBN 0-387-94858-9

- Also appears as Heckerman, David (March 1997). "Bayesian Networks for Data Mining". Data Mining and Knowledge Discovery. 1 (1): 79–119. doi:10.1023/A:1009730122752

- Andrew Gelman; John B Carlin; Hal S Stern; Donald B Rubin (2003). "Part II: Fundamentals of Bayesian Data Analysis: Ch.5 Hierarchical models". Bayesian Data Analysis. CRC Press. pp. 120–. ISBN 978-1-58488-388-3

- Jøsang, A. and McAnally, D., 2004, "Multiplication and Comultiplication of Beliefs," International Journal of Approximate Reasoning, 38(1), pp.19-51, 2004

- Jensen, Finn V; Nielsen, Thomas D. (June 6, 2007). Bayesian Networks and Decision Graphs. Information Science and Statistics series (2nd ed.). New York: Springer-Verlag. ISBN 978-0-387-68281-5

- Lunn, David; Spiegelhalter, David; Thomas, Andrew; Best, Nicky; et al. (November 2009). "The BUGS project: Evolution, critique and future directions". Statistics in Medicine. 28 (25): 3049–3067. PMID 19630097. doi:10.1002/sim.3680

- Korb, Kevin B.; Nicholson, Ann E. (December 2010). Bayesian Artificial Intelligence. CRC Computer Science & Data Analysis (2nd ed.). Chapman & Hall (CRC Press). doi:10.1007/s10044-004-0214-5. ISBN 1-58488-387-1

- Jøsang, A., 2008, "Conditional Reasoning with Subjective Logic," Journal of Multiple-Valued Logic and Soft Computing, 15(1), pp.5-38, 2008

- Pearl, Judea (1988). Probabilistic Reasoning in Intelligent Systems: Networks of Plausible Inference. Representation and Reasoning Series (2nd printing ed.). San Francisco, California: Morgan Kaufmann. ISBN 0-934613-73-7

- Pearl, Judea (September 1986). "Fusion, propagation, and structuring in belief networks". Artificial Intelligence. 29 (3): 241–288. doi:10.1016/0004-3702(86)90072-X

- Aster, Richard; Borchers, Brian, and Thurber, Clifford (2012). Parameter Estimation and Inverse Problems, Second Edition, Elsevier. ISBN 0123850487, ISBN 978-0123850485

- Ruspini, E.H., Lowrance, J., and Strat, T., 1992, "Understanding evidential reasoning," International Journal of Approximate Reasoning, 6(3): 401-424

- Bickel, Peter J. & Doksum, Kjell A. (2001). Mathematical Statistics, Volume 1: Basic and Selected Topics (Second (updated printing 2007) ed.). Pearson Prentice–Hall. ISBN 0-13-850363-X

Machine Learning: An Overview

Machine learning is an interdisciplinary branch that deals with creating a machine that can learn with experience and not just by programming. Unlike classical artificial intelligence that uses deductive reasoning, this method follows inductive reasoning. While deductive reasoning solves a problem on the basis of general axioms, inductive reasoning takes examples as the base and generates general axioms. The major concepts of machine learning are discussed in this section.

Machine Learning

Machine Learning is the study of how to build computer systems that adapt and improve with experience. It is a subfield of Artificial Intelligence and intersects with cognitive science, information theory, and probability theory, among others.

Classical AI deals mainly with *deductive* reasoning, learning represents *inductive* reasoning. Deductive reasoning arrives at answers to queries relating to a particular situation starting from a set of general axioms, whereas inductive reasoning arrives at general axioms from a set of particular instances.

Classical AI often suffers from the knowledge acquisition problem in real life applications where obtaining and updating the knowledge base is costly and prone to errors. Machine learning serves to solve the knowledge acquisition bottleneck by obtaining the result from data by induction.

Machine learning is particularly attractive in several real life problem because of the following reasons:

- Some tasks cannot be defined well except by example
- Working environment of machines may not be known at design time
- Explicit knowledge encoding may be difficult and not available
- Environments change over time
- Biological systems learn

Recently, learning is widely used in a number of application areas including,

- Data mining and knowledge discovery
- Speech/image/video (pattern) recognition
- Adaptive control
- Autonomous vehicles/robots

- Decision support systems
- Bioinformatics
- WWW

Formally, a computer program is said to learn from experience E with respect to some class of tasks T and performance measure P, if its performance at tasks in T, as measured by P, improves with experience E.

Thus a learning system is characterized by:

- task T
- experience E, and
- performance measure P

Examples:

Learning to play chess

T: Play chess

P: Percentage of games won in world tournament

E: Opportunity to play against self or other players

Learning to drive a van

T: Drive on a public highway using vision sensors

P: Average distance traveled before an error (according to human observer)

E: Sequence of images and steering actions recorded during human driving.

The block diagram of a generic learning system which can realize the above definition is shown below:

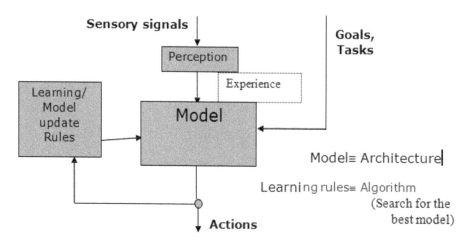

As can be seen from the above diagram the system consists of the following components:

- *Goal:* Defined with respect to the task to be performed by the system

- *Model:* A mathematical function which maps perception to actions

- *Learning rules:* Which update the model parameters with new experience such that the performance measures with respect to the goals is optimized

- *Experience:* A set of perception (and possibly the corresponding actions)

Taxonomy of Learning Systems:

Several classification of learning systems are possible based on the above components as follows:

Goal/Task/Target Function:

Prediction: To predict the desired output for a given input based on previous input/output pairs. E.g., to predict the value of a stock given other inputs like market index, interest rates etc.

Categorization: To classify an object into one of several categories based on features of the object. E.g., a robotic vision system to categorize a machine part into one of the categories, spanner, hammer etc based on the parts' dimension and shape.

Clustering: To organize a group of objects into homogeneous segments. E.g., a satellite image analysis system which groups land areas into forest, urban and water body, for better utilization of natural resources.

Planning: To generate an optimal sequence of actions to solve a particular problem. E.g., an Unmanned Air Vehicle which plans its path to obtain a set of pictures and avoid enemy anti-aircraft guns.

Models:

- Propositional and FOL rules

- Decision trees

- Linear separators

- Neural networks

- Graphical models

- Temporal models like hidden Markov models

Learning Rules:

Learning rules are often tied up with the model of learning used. Some common rules are gradient descent, least square error, expectation maximization and margin maximization.

Experiences:

Learning algorithms use experiences in the form of perceptions or perception action pairs to improve their performance. The nature of experiences available varies with applications. Some common situations are described below.

Supervised learning: In supervised learning a teacher or oracle is available which provides the desired action corresponding to a perception. A set of perception action pair provides what is called a training set. Examples include an automated vehicle where a set of vision inputs and the corresponding steering actions are available to the learner.

Unsupervised learning: In unsupervised learning no teacher is available. The learner only discovers persistent patterns in the data consisting of a collection of perceptions. This is also called exploratory learning. Finding out malicious network attacks from a sequence of anomalous data packets is an example of unsupervised learning.

Active learning: Here not only a teacher is available, the learner has the freedom to ask the teacher for suitable perception-action example pairs which will help the learner to improve its performance. Consider a news recommender system which tries to learn an users preferences and categorize news articles as interesting or uninteresting to the user. The system may present a particular article (of which it is not sure) to the user and ask whether it is interesting or not.

Reinforcement learning: In reinforcement learning a teacher is available, but the teacher instead of directly providing the desired action corresponding to a perception, return reward and punishment to the learner for its action corresponding to a perception.

Examples include a robot in a unknown terrain where its get a punishment when its hits an obstacle and reward when it moves smoothly.

In order to design a learning system the designer has to make the following choices based on the application.

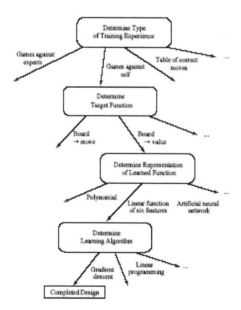

Mathematical formulation of the inductive learning problem:

- Extrapolate from a given set of examples so that we can make accurate predictions about future examples.

- Supervised versus Unsupervised learning

Want to learn an unknown function $f(x) = y$, where x is an input example and y is the desired output. Supervised learning implies we are given a set of (x, y) pairs by a "teacher." Unsupervised learning means we are only given the xs. In either case, the goal is to estimate f.

Inductive Bias:

- Inductive learning is an inherently conjectural process because any knowledge created by generalization from specific facts cannot be proven true; it can only be proven false. Hence, inductive inference is falsity preserving, not truth preserving.

- To generalize beyond the specific training examples, we need constraints or biases on what f is best. That is, learning can be viewed as searching the Hypothesis Space H of possible f functions.

- A bias allows us to choose one f over another one

- A completely unbiased inductive algorithm could only memorize the training examples and could not say anything more about other unseen examples.

- Two types of biases are commonly used in machine learning:

  - Restricted Hypothesis Space Bias

Allow only certain types of f functions, not arbitrary ones

  - Preference Bias

Define a metric for comparing fs so as to determine whether one is better than another

Inductive Learning Framework:

- Raw input data from sensors are preprocessed to obtain a feature vector, x, that adequately describes all of the relevant features for classifying examples.

- Each x is a list of (attribute, value) pairs. For example,

x = (Person = Sue, Eye-Color = Brown, Age = Young, Sex = Female)

The number of attributes (also called features) is fixed (positive, finite). Each attribute has a fixed, finite number of possible values.

Each example can be interpreted as a *point* in an n-dimensional feature space, where n is the number of attributes.

Machine learning is the subfield of computer science that, according to Arthur Samuel in 1959, gives "computers the ability to learn without being explicitly programmed." Evolved from the study of pattern recognition and computational learning theory in artificial intelligence, machine learning explores the study and construction of algorithms that can learn from and make predictions on data – such algorithms overcome following strictly static program instructions by making data-driven predictions or decisions, through building a model from sample inputs. Machine learning is employed in a range of computing tasks where designing and programming explicit algorithms with good performance is difficult or unfeasible; example applications include email filtering, detection of network intruders or malicious insiders working towards a data breach, optical character recognition (OCR), learning to rank and computer vision.

Machine learning is closely related to (and often overlaps with) computational statistics, which also focuses on prediction-making through the use of computers. It has strong ties to mathematical optimization, which delivers methods, theory and application domains to the field. Machine learning is sometimes conflated with data mining, where the latter subfield focuses more on exploratory data analysis and is known as unsupervised learning.[vii] Machine learning can also be unsupervised and be used to learn and establish baseline behavioral profiles for various entities and then used to find meaningful anomalies.

Within the field of data analytics, machine learning is a method used to devise complex models and algorithms that lend themselves to prediction; in commercial use, this is known as predictive analytics. These analytical models allow researchers, data scientists, engineers, and analysts to "produce reliable, repeatable decisions and results" and uncover "hidden insights" through learning from historical relationships and trends in the data.

As of 2016, machine learning is a buzzword, and according to the Gartner hype cycle of 2016, at its peak of inflated expectations. Because finding patterns is hard, often not enough training data is available, and also because of the high expectations it often fails to deliver.

Overview

Tom M. Mitchell provided a widely quoted, more formal definition: "A computer program is said to learn from experience E with respect to some class of tasks T and performance measure P if its performance at tasks in T, as measured by P, improves with experience E." This definition is notable for its defining machine learning in fundamentally operational rather than cognitive terms, thus following Alan Turing's proposal in his paper "Computing Machinery and Intelligence", that the question "Can machines think?" be replaced with the question "Can machines do what we (as thinking entities) can do?". In the proposal he explores the various characteristics that could be possessed by a *thinking machine* and the various implications in constructing one.

Types of Problems and Tasks

Machine learning tasks are typically classified into three broad categories, depending on the nature of the learning "signal" or "feedback" available to a learning system. These are

- Supervised learning: The computer is presented with example inputs and their desired outputs, given by a "teacher", and the goal is to learn a general rule that maps inputs to outputs.

- Unsupervised learning: No labels are given to the learning algorithm, leaving it on its own to find structure in its input. Unsupervised learning can be a goal in itself (discovering hidden patterns in data) or a means towards an end (feature learning).

- Reinforcement learning: A computer program interacts with a dynamic environment in which it must perform a certain goal (such as driving a vehicle or playing a game against an opponent[3]). The program is provided feedback in terms of rewards and punishments as it navigates its problem space.

Between supervised and unsupervised learning is semi-supervised learning, where the teacher gives an incomplete training signal: a training set with some (often many) of the target outputs missing. Transduction is a special case of this principle where the entire set of problem instances is known at learning time, except that part of the targets are missing.

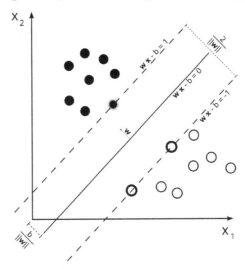

A support vector machine is a classifier that divides its input space into two regions, separated by a linear boundary. Here, it has learned to distinguish black and white circles.

Among other categories of machine learning problems, learning to learn learns its own inductive bias based on previous experience. Developmental learning, elaborated for robot learning, generates its own sequences (also called curriculum) of learning situations to cumulatively acquire repertoires of novel skills through autonomous self-exploration and social interaction with human teachers and using guidance mechanisms such as active learning, maturation, motor synergies, and imitation.

Another categorization of machine learning tasks arises when one considers the desired *output* of a machine-learned system:

- In classification, inputs are divided into two or more classes, and the learner must produce a model that assigns unseen inputs to one or more (multi-label classification) of these classes. This is typically tackled in a supervised way. Spam filtering is an example of classification, where the inputs are email (or other) messages and the classes are "spam" and "not spam".

- In regression, also a supervised problem, the outputs are continuous rather than discrete.

- In clustering, a set of inputs is to be divided into groups. Unlike in classification, the groups are not known beforehand, making this typically an unsupervised task.

- Density estimation finds the distribution of inputs in some space.

- Dimensionality reduction simplifies inputs by mapping them into a lower-dimensional space. Topic modeling is a related problem, where a program is given a list of human language documents and is tasked to find out which documents cover similar topics.

History and Relationships to other Fields

As a scientific endeavour, machine learning grew out of the quest for artificial intelligence. Already in the early days of AI as an academic discipline, some researchers were interested in having machines learn from data. They attempted to approach the problem with various symbolic methods, as well as what were then termed "neural networks"; these were mostly perceptrons and other models that were later found to be reinventions of the generalized linear models of statistics. Probabilistic reasoning was also employed, especially in automated medical diagnosis.

However, an increasing emphasis on the logical, knowledge-based approach caused a rift between AI and machine learning. Probabilistic systems were plagued by theoretical and practical problems of data acquisition and representation. By 1980, expert systems had come to dominate AI, and statistics was out of favor. Work on symbolic/knowledge-based learning did continue within AI, leading to inductive logic programming, but the more statistical line of research was now outside the field of AI proper, in pattern recognition and information retrieval. Neural networks research had been abandoned by AI and computer science around the same time. This line, too, was continued outside the AI/CS field, as "connectionism", by researchers from other disciplines including Hopfield, Rumelhart and Hinton. Their main success came in the mid-1980s with the reinvention of backpropagation.

Machine learning, reorganized as a separate field, started to flourish in the 1990s. The field changed its goal from achieving artificial intelligence to tackling solvable problems of a practical nature. It shifted focus away from the symbolic approaches it had inherited from AI, and toward methods and models borrowed from statistics and probability theory. It also benefited from the increasing availability of digitized information, and the possibility to distribute that via the Internet.

Machine learning and data mining often employ the same methods and overlap significantly, but while machine learning focuses on prediction, based on *known* properties learned from the training data, data mining focuses on the discovery of (previously) *unknown* properties in the data (this is the analysis step of Knowledge Discovery in Databases). Data mining uses many machine learning methods, but with different goals; on the other hand, machine learning also employs data mining methods as "unsupervised learning" or as a preprocessing step to improve learner accuracy. Much of the confusion between these two research communities (which do often have separate conferences and separate journals, ECML PKDD being a major exception) comes from the basic assumptions they work with: in machine learning, performance is usually evaluated with respect to the ability to *reproduce known* knowledge, while in Knowledge Discovery and Data Mining (KDD) the key task is the discovery of previously *unknown* knowledge. Evaluated with respect to known knowledge, an uninformed (unsupervised) method will easily be outperformed by other supervised methods, while in a typical KDD task, supervised methods cannot be used due to the unavailability of training data.

Machine learning also has intimate ties to optimization: many learning problems are formulated as minimization of some loss function on a training set of examples. Loss functions express the discrepancy between the predictions of the model being trained and the actual problem instances (for example, in classification, one wants to assign a label to instances, and models are trained to correctly predict the pre-assigned labels of a set examples). The difference between the two fields arises from the goal of generalization: while optimization algorithms can minimize the loss on a training set, machine learning is concerned with minimizing the loss on unseen samples.

Relation to Statistics

Machine learning and statistics are closely related fields. According to Michael I. Jordan, the ideas of machine learning, from methodological principles to theoretical tools, have had a long pre-history in statistics. He also suggested the term data science as a placeholder to call the overall field.

Leo Breiman distinguished two statistical modelling paradigms: data model and algorithmic model, wherein 'algorithmic model' means more or less the machine learning algorithms like Random forest.

Some statisticians have adopted methods from machine learning, leading to a combined field that they call *statistical learning*.

Theory

A core objective of a learner is to generalize from its experience. Generalization in this context is the ability of a learning machine to perform accurately on new, unseen examples/tasks after having experienced a learning data set. The training examples come from some generally unknown probability distribution (considered representative of the space of occurrences) and the learner has to build a general model about this space that enables it to produce sufficiently accurate predictions in new cases.

The computational analysis of machine learning algorithms and their performance is a branch of theoretical computer science known as computational learning theory. Because training sets are finite and the future is uncertain, learning theory usually does not yield guarantees of the performance of algorithms. Instead, probabilistic bounds on the performance are quite common. The bias–variance decomposition is one way to quantify generalization error.

For the best performance in the context of generalization, the complexity of the hypothesis should match the complexity of the function underlying the data. If the hypothesis is less complex than the function, then the model has underfit the data. If the complexity of the model is increased in response, then the training error decreases. But if the hypothesis is too complex, then the model is subject to overfitting and generalization will be poorer.

In addition to performance bounds, computational learning theorists study the time complexity and feasibility of learning. In computational learning theory, a computation is considered feasible if it can be done in polynomial time. There are two kinds of time complexity results. Positive results show that a certain class of functions can be learned in polynomial time. Negative results show that certain classes cannot be learned in polynomial time.

Approaches

Decision Tree Learning

Decision tree learning uses a decision tree as a predictive model, which maps observations about an item to conclusions about the item's target value.

Association Rule Learning

Association rule learning is a method for discovering interesting relations between variables in large databases.

Artificial Neural Networks

An artificial neural network (ANN) learning algorithm, usually called "neural network" (NN), is a learning algorithm that is inspired by the structure and functional aspects of biological neural networks. Computations are structured in terms of an interconnected group of artificial neurons, processing information using a connectionist approach to computation. Modern neural networks are non-linear statistical data modeling tools. They are usually used to model complex relationships between inputs and outputs, to find patterns in data, or to capture the statistical structure in an unknown joint probability distribution between observed variables.

Deep Learning

Falling hardware prices and the development of GPUs for personal use in the last few years have contributed to the development of the concept of Deep learning which consists of multiple hidden layers in an artificial neural network. This approach tries to model the way the human brain processes light and sound into vision and hearing. Some successful applications of deep learning are computer vision and speech recognition.

Inductive Logic Programming

Inductive logic programming (ILP) is an approach to rule learning using logic programming as a uniform representation for input examples, background knowledge, and hypotheses. Given an encoding of the known background knowledge and a set of examples represented as a logical database of facts, an ILP system will derive a hypothesized logic program that entails all positive and no negative examples. Inductive programming is a related field that considers any kind of programming languages for representing hypotheses (and not only logic programming), such as functional programs.

Support Vector Machines

Support vector machines (SVMs) are a set of related supervised learning methods used for classification and regression. Given a set of training examples, each marked as belonging to one of two categories, an SVM training algorithm builds a model that predicts whether a new example falls into one category or the other.

Clustering

Cluster analysis is the assignment of a set of observations into subsets (called *clusters*) so that observations within the same cluster are similar according to some predesignated criterion or criteria, while observations drawn from different clusters are dissimilar. Different clustering techniques make different assumptions on the structure of the data, often defined by some *similarity metric* and evaluated for example by *internal compactness* (similarity between members of the same cluster) and *separation* between different clusters. Other methods are based on *estimated density* and *graph connectivity*. Clustering is a method of unsupervised learning, and a common technique for statistical data analysis.

Bayesian Networks

A Bayesian network, belief network or directed acyclic graphical model is a probabilistic graphical model that represents a set of random variables and their conditional independencies via a directed acyclic graph (DAG). For example, a Bayesian network could represent the probabilistic relationships between diseases and symptoms. Given symptoms, the network can be used to compute the probabilities of the presence of various diseases. Efficient algorithms exist that perform inference and learning.

Reinforcement Learning

Reinforcement learning is concerned with how an *agent* ought to take *actions* in an *environment* so as to maximize some notion of long-term *reward*. Reinforcement learning algorithms attempt to find a *policy* that maps *states* of the world to the actions the agent ought to take in those states. Reinforcement learning differs from the supervised learning problem in that correct input/output pairs are never presented, nor sub-optimal actions explicitly corrected.

Representation Learning

Several learning algorithms, mostly unsupervised learning algorithms, aim at discovering better representations of the inputs provided during training. Classical examples include principal components analysis and cluster analysis. Representation learning algorithms often attempt to preserve the information in their input but transform it in a way that makes it useful, often as a pre-processing step before performing classification or predictions, allowing reconstruction of the inputs coming from the unknown data generating distribution, while not being necessarily faithful for configurations that are implausible under that distribution.

Manifold learning algorithms attempt to do so under the constraint that the learned representation is low-dimensional. Sparse coding algorithms attempt to do so under the constraint that the learned representation is sparse (has many zeros). Multilinear subspace learning algorithms aim to learn low-dimensional representations directly from tensor representations for multidimensional data, without reshaping them into (high-dimensional) vectors. Deep learning algorithms discover multiple levels of representation, or a hierarchy of features, with higher-level, more abstract features defined in terms of (or generating) lower-level features. It has been argued that an intelligent machine is one that learns a representation that disentangles the underlying factors of variation that explain the observed data.

Similarity and Metric Learning

In this problem, the learning machine is given pairs of examples that are considered similar and pairs of less similar objects. It then needs to learn a similarity function (or a distance metric function) that can predict if new objects are similar. It is sometimes used in Recommendation systems.

Sparse Dictionary Learning

In this method, a datum is represented as a linear combination of basis functions, and the coefficients are assumed to be sparse. Let x be a d-dimensional datum, D be a d by n matrix, where each column of D represents a basis function. r is the coefficient to represent x using D. Mathematically, sparse dictionary learning means solving $x \approx Dr$ where r is sparse. Generally speaking, n is assumed to be larger than d to allow the freedom for a sparse representation.

Learning a dictionary along with sparse representations is strongly NP-hard and also difficult to solve approximately. A popular heuristic method for sparse dictionary learning is K-SVD.

Sparse dictionary learning has been applied in several contexts. In classification, the problem is to determine which classes a previously unseen datum belongs to. Suppose a dictionary for each class has already been built. Then a new datum is associated with the class such that it's best sparsely represented by the corresponding dictionary. Sparse dictionary learning has also been applied in image de-noising. The key idea is that a clean image patch can be sparsely represented by an image dictionary, but the noise cannot.

Genetic Algorithms

A genetic algorithm (GA) is a search heuristic that mimics the process of natural selection, and uses methods such as mutation and crossover to generate new genotype in the hope of finding good solutions to a given problem. In machine learning, genetic algorithms found some uses in the 1980s and 1990s. Vice versa, machine learning techniques have been used to improve the performance of genetic and evolutionary algorithms.

Rule-based Machine Learning

Rule-based machine learning is a general term for any machine learning method that identifies, learns, or evolves `rules' to store, manipulate or apply, knowledge. The defining characteristic of a rule-based machine learner is the identification and utilization of a set of relational rules that collectively represent the knowledge captured by the system. This is in contrast to other machine learners that commonly identify a singular model that can be universally applied to any instance in order to make a prediction. Rule-based machine learning approaches include learning classifier systems, association rule learning, and artificial immune systems.

Learning Classifier Systems

Learning classifier systems (LCS) are a family of rule-based machine learning algorithms that combine a discovery component (e.g. typically a genetic algorithm) with a learning component (performing either supervised learning, reinforcement learning, or unsupervised learning). They seek to identify a set of context-dependent rules that collectively store and apply knowledge in a piecewise manner in order to make predictions.

Applications

Applications for machine learning include:

- Adaptive websites

- Affective computing

- Bioinformatics

- Brain-machine interfaces

- Cheminformatics

- Classifying DNA sequences

- Computational anatomy

- Computer vision, including object recognition

- Detecting credit card fraud

- Game playing

- Information retrieval

- Internet fraud detection

- Marketing

- Machine learning control

- Machine perception

- Medical diagnosis

- Economics

- Natural language processing

- Natural language understanding

- Optimization and metaheuristic

- Online advertising

- Recommender systems

- Robot locomotion

- Search engines

- Sentiment analysis (or opinion mining)

- Sequence mining

- Software engineering

- Speech and handwriting recognition

- Financial market analysis

- Structural health monitoring

- Syntactic pattern recognition

- User behavior analytics

- Translation

In 2006, the online movie company Netflix held the first "Netflix Prize" competition to find a program to better predict user preferences and improve the accuracy on its existing Cinematch movie recommendation algorithm by at least 10%. A joint team made up of researchers from AT&T Labs-Research in collaboration with the teams Big Chaos and Pragmatic Theory built an ensemble model to win the Grand Prize in 2009 for $1 million. Shortly after the prize was awarded, Netflix realized that viewers' ratings were not the best indicators of their viewing patterns ("everything is a recommendation") and they changed their recommendation engine accordingly.

In 2010 The Wall Street Journal wrote about money management firm Rebellion Research's use of machine learning to predict economic movements. The article describes Rebellion Research's prediction of the financial crisis and economic recovery.

In 2012 co-founder of Sun Microsystems Vinod Khosla predicted that 80% of medical doctors jobs would be lost in the next two decades to automated machine learning medical diagnostic software.

In 2014 it has been reported that a machine learning algorithm has been applied in Art History to study fine art paintings, and that it may have revealed previously unrecognized influences between artists.

Model Assessments

Classification machine learning models can be validated by accuracy estimation techniques like the Holdout method, which splits the data in a training and test set (conventionally 2/3 training set and 1/3 test set designation) and evaluates the performance of the training model on the test set. In comparison, the N-fold-cross-validation method randomly splits the data in k subsets where the k-1 instances of the data are used to train the model while the kth instance is used to test the predictive ability of the training model. In addition to the holdout and cross-validation methods, bootstrap, which samples n instances with replacement from the dataset, can be used to assess model accuracy. In addition to accuracy, sensitivity and specificity (True Positive Rate: TPR and True Negative Rate: TNR, respectively) can provide modes of model assessment. Similarly False Positive Rate (FPR) as well as the False Negative Rate (FNR) can be computed. Receiver operating characteristic (ROC) along with the accompanying Area Under the ROC Curve (AUC) offer additional tools for classification model assessment. Higher AUC is associated with a better performing model.

Ethics

Machine Learning poses a host of ethical questions. Systems which are trained on datasets collected with biases may exhibit these biases upon use, thus digitizing cultural prejudices. Responsible collection of data thus is a critical part of machine learning.

Because language contains biases, machines trained on language corpora will necessarily also learn bias.

Software

Software suites containing a variety of machine learning algorithms include the following :

Free and Open-source Software

- dlib
- ELKI
- Encog
- GNU Octave
- H2O
- Mahout
- Mallet (software project)
- mlpy
- MLPACK
- MOA (Massive Online Analysis)
- ND4J with Deeplearning4j
- NuPIC
- OpenAI Gym
- OpenAI Universe
- OpenNN
- Orange
- Python
- R
- scikit-learn
- Shogun
- TensorFlow
- Torch (machine learning)
- Spark

- Yooreeka

- Weka

Proprietary Software with Free and Open-source Editions

- KNIME

- RapidMiner

Proprietary Software

- Amazon Machine Learning

- Angoss KnowledgeSTUDIO

- Ayasdi

- Databricks

- Google Prediction API

- IBM SPSS Modeler

- KXEN Modeler

- LIONsolver

- Mathematica

- MATLAB

- Microsoft Azure Machine Learning

- Neural Designer

- NeuroSolutions

- Oracle Data Mining

- RCASE

- SAS Enterprise Miner

- SequenceL

- Splunk

- STATISTICA Data Miner

Concept Learning

Definition:

The problem is to learn a function mapping examples into two classes: positive and negative. We are given a database of examples already classified as positive or negative. Concept learning: the process of inducing a function mapping input examples into a Boolean output.

Examples:

- Classifying objects in astronomical images as stars or galaxies

- Classifying animals as vertebrates or invertebrates

Example: Classifying Mushrooms

Class of Tasks:	Predicting poisonous mushrooms
Performance:	Accuracy of classification
Experience:	Database describing mushrooms with their class
Knowledge to learn:	Function mapping mushrooms to {0,1} where 0:not-poisonous and 1:poisonous

Representation of target knowledge: conjunction of attribute values. Learning mechanism: candidate-elimination

Representation of instances:

Features:

- color   {red, brown, gray}

- size   {small, large}

- shape   {round,elongated}

- land   {humid,dry}

- air humidity   {low,high}

- texture   {smooth, rough}

Input and Output Spaces:

X : The space of all possible examples (input space).

Y: The space of classes (output space).

An example in X is a feature vector X.

> For instance: X = (red,small,elongated,humid,low,rough) X is the cross product of all feature values.

Only a small subset of instances is available in the database of examples.

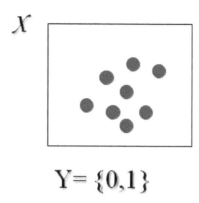

$$X$$

$$Y = \{0,1\}$$

Training Examples:

D : The set of training examples.

D is a set of pairs { (x,c(x)) }, where c is the target concept. c is a subset of the universe of discourse or the set of all possible instances.

Example of D:

((red,small,round,humid,low,smooth),   poisonous)

((red,small,elongated,humid,low,smooth),   poisonous)

((gray,large,elongated,humid,low,rough),   not-poisonous)

((red,small,elongated,humid,high,rough),   poisonous)

Hypothesis Representation

Any hypothesis h is a function from X to Y

h: X → Y

We will explore the space of conjunctions.

Special symbols:

> ? Any value is acceptable

> o no value is acceptable

Consider the following hypotheses:

(?,?,?,?,?,?): all mushrooms are poisonous

(o,o,o,o,o,o): no mushroom is poisonous

Hypotheses Space:

The space of all hypotheses is represented by H Let h be a hypothesis in H.

Let X be an example of a mushroom.

if h(X) = 1 then X is poisonous, otherwise X is not-poisonous

Our goal is to find the hypothesis, h*, that is very "close" to target concept c.

A hypothesis is said to "cover" those examples it classifies as positive.

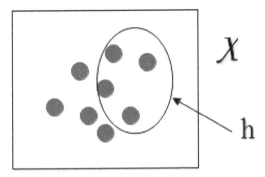

Assumptions:

We will explore the space of all conjunctions.

We assume the target concept falls within this space.

A hypothesis close to target concept c obtained after seeing many training examples will result in high accuracy on the set of unobserved examples. (Inductive Learning Hypothesis)

Concept Learning as Search

We will see how the problem of concept learning can be posed as a search problem.

We will illustrate that there is a general to specific ordering inherent to any hypothesis space.

Consider these two hypotheses: h1 = (red,?,?,humid,?,?)

h2 = (red,?,?,?,?,?)

We say h2 is more general than h1 because h2 classifies more instances than h1 and h1 is covered by h2.

For example, consider the following hypotheses

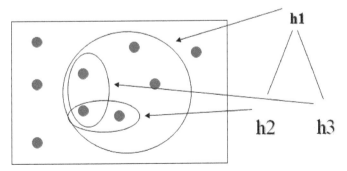

h1 is more general than h2 and h3.

h2 and h3 are neither more specific nor more general than each other.

Definitions:

Let hj and hk be two hypotheses mapping examples into {0,1}. We say hj is more general than hk iff

For all examples X, hk(X) = 1 -7 hj(X) = 1

We represent this fact as hj >= hk

The >= relation imposes a partial ordering over the hypothesis space H (reflexive, antisymmetric, and transitive).

Any input space X defines then a lattice of hypotheses ordered according to the general- specific relation:

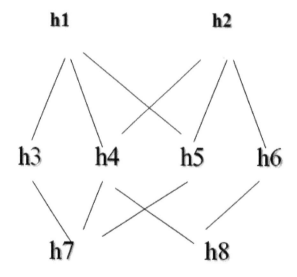

Algorithm to Find a Maximally-Specific Hypothesis

Algorithm to search the space of conjunctions:

- Start with the most specific hypothesis

- Generalize the hypothesis when it fails to cover a positive example

Algorithm:

1. Initialize h to the most specific hypothesis

2. For each positive training example X

   For each value a in h

   If example X and h agree on a, do nothing

   Else generalize a by the next more general constraint

1. Output hypothesis h

Example:

Let's run the learning algorithm above with the following examples:

((red,small,round,humid,low,smooth),   poisonous)

((red,small,elongated,humid,low,smooth),   poisonous)

((gray,large,elongated,humid,low,rough),   not-poisonous)

((red,small,elongated,humid,high,rough),   poisonous)

We start with the most specific hypothesis: h = (0,0,0,0,0,0)

The first example comes and since the example is positive and h fails to cover it, we simply generalize h to cover exactly this example:

h = (red,small,round,humid,low,smooth)

Hypothesis h basically says that the first example is the only positive example, all other examples are negative.

Then comes examples 2: ((red,small,elongated,humid,low,smooth), poisonous)

This example is positive. All attributes match hypothesis h except for attribute shape: it has the value *elongated*, not *round*. We generalize this attribute using symbol ? yielding:

h: (red,small,?,humid,low,smooth)

The third example is negative and so we just ignore it.

Why is it we don't need to be concerned with negative examples?

Upon observing the 4th example, hypothesis h is generalized to the following:

h = (red,small,?,humid,?,?)

h is interpreted as any mushroom that is red, small and found on humid land should be classified as poisonous.

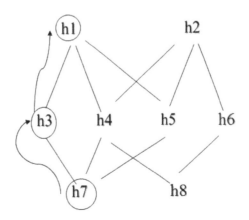

The algorithm is guaranteed to find the hypothesis that is most specific and consistent with the set of training examples. It takes advantage of the general-specific ordering to move on the corresponding lattice searching for the next most specific hypothesis.

Note that:

There are many hypotheses consistent with the training data D. Why should we prefer the most specific hypothesis?

What would happen if the examples are not consistent? What would happen if they have errors, noise?

What if there is a hypothesis space H where one can find more that one maximally specific hypothesis h? The search over the lattice must then be different to allow for this possibility.

- The algorithm that finds the maximally specific hypothesis is limited in that it only finds one of many hypotheses consistent with the training data.

- The Candidate Elimination Algorithm (CEA) finds ALL hypotheses consistent with the training data.

- CEA does that without explicitly enumerating all consistent hypotheses.

- Applications:

  - Chemical Mass Spectroscopy

  - Control Rules for Heuristic Search

Candidate Elimination Algorithm

Consistency vs Coverage

In the following example, h1 covers a different set of examples than h2, h2 is consistent with training set D, h1 is not consistent with training set D

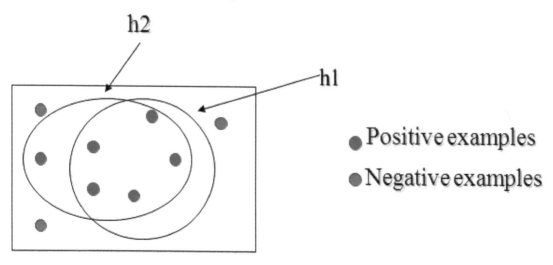

Version Space

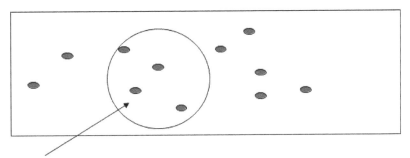

Hypothesis space $\mathcal{H}$

Version space: Subset of hypothesis from $\mathcal{H}$ consistent with training set $\mathcal{D}$.

The version space for the mushroom example is as follows:

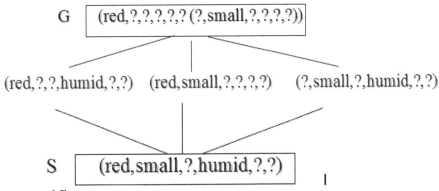

S: Most specific

G: Most general

The candidate elimination algorithm generates the entire version space.

The Candidate-Elimination Algorithm

The candidate elimination algorithm keeps two lists of hypotheses consistent with the training data: (i) The list of most specific hypotheses S and, (ii) The list of most general hypotheses G. This is enough to derive the whole version space VS.

Steps:

1. Initialize G to the set of maximally general hypotheses in H

2. Initialize S to the set of maximally specific hypotheses in H

3. For each training example X do

    (a) If X is positive: generalize S if necessary

    (b) If X is negative: specialize G if necessary

4. Output {G,S}

Step (a) Positive examples

If X is positive:

- Remove from G any hypothesis inconsistent with X

- For each hypothesis h in S not consistent with X

  - Remove h from S

  - Add all minimal generalizations of h consistent with X such that some member of G is more general than h

  - Remove from S any hypothesis more general than any other hypothesis in S

Step (b) Negative examples

If X is negative:

- Remove from S any hypothesis inconsistent with X

- For each hypothesis h in G not consistent with X

  - Remove g from G

  - Add all minimal generalizations of h consistent with X such that some member of S is more specific than h

  - Remove from G any hypothesis less general than any other hypothesis in G

The candidate elimination algorithm is guaranteed to converge to the right hypothesis provided the following:

a)   No errors exist in the examples

b)   The target concept is included in the hypothesis space H

If there exists errors in the examples:

a)   The right hypothesis would be inconsistent and thus eliminated.

b)   If the S and G sets converge to an empty space we have evidence that the true concept lies outside space H.

Decision Tree

A decision tree is a decision support tool that uses a tree-like graph or model of decisions and their possible consequences, including chance event outcomes, resource costs, and utility. It is one way to display an algorithm.

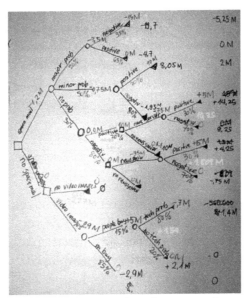

Traditionally, decision trees have been created manually.

Decision trees are commonly used in operations research, specifically in decision analysis, to help identify a strategy most likely to reach a goal, but are also a popular tool in machine learning.

Overview

A decision tree is a flowchart-like structure in which each internal node represents a "test" on an attribute (e.g. whether a coin flip comes up heads or tails), each branch represents the outcome of the test, and each leaf node represents a class label (decision taken after computing all attributes). The paths from root to leaf represent classification rules.

In decision analysis, a decision tree and the closely related influence diagram are used as a visual and analytical decision support tool, where the expected values (or expected utility) of competing alternatives are calculated.

A decision tree consists of three types of nodes:

1. Decision nodes – typically represented by squares

2. Chance nodes – typically represented by circles

3. End nodes – typically represented by triangles

Decision trees are commonly used in operations research and operations management. If, in practice, decisions have to be taken online with no recall under incomplete knowledge, a decision tree should be paralleled by a probability model as a best choice model or online selection model algorithm. Another use of decision trees is as a descriptive means for calculating conditional probabilities.

Decision trees, influence diagrams, utility functions, and other decision analysis tools and methods are taught to undergraduate students in schools of business, health economics, and public health, and are examples of operations research or management science methods.

Decision Tree Building Blocks

Decision Tree Elements

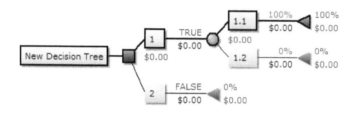

Drawn from left to right, a decision tree has only burst nodes (splitting paths) but no sink nodes (converging paths). Therefore, used manually, they can grow very big and are then often hard to draw fully by hand. Traditionally, decision trees have been created manually — as the example shows — although increasingly, specialized software is employed.

Decision Rules

The decision tree can be linearized into decision rules, where the outcome is the contents of the leaf node, and the conditions along the path form a conjunction in the if clause. In general, the rules have the form:

> *if* condition1 *and* condition2 *and* condition3 *then* outcome.

Decision rules can be generated by constructing association rules with the target variable on the right. They can also denote temporal or causal relations.

Decision Tree using Flowchart Symbols

Commonly a decision tree is drawn using flowchart symbols as it is easier for many to read and understand.

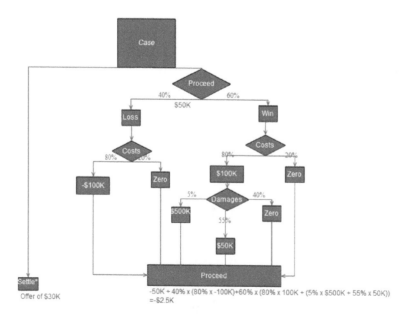

Analysis Example

Analysis can take into account the decision maker's (e.g., the company's) preference or utility function, for example:

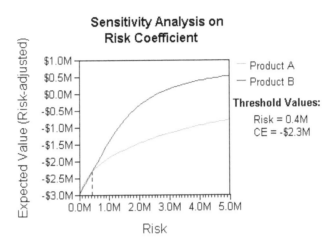

The basic interpretation in this situation is that the company prefers B's risk and payoffs under realistic risk preference coefficients (greater than $400K—in that range of risk aversion, the company would need to model a third strategy, "Neither A nor B").

Influence Diagram

Much of the information in a decision tree can be represented more compactly as an influence diagram, focusing attention on the issues and relationships between events.

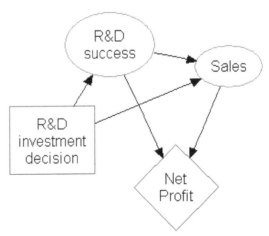

The rectangle on the left represents a decision, the ovals represent actions, and the diamond represents results.

Association Rule Induction

Decision trees can also be seen as generative models of induction rules from empirical data. An optimal decision tree is then defined as a tree that accounts for most of the data, while minimizing the number of levels (or "questions"). Several algorithms to generate such optimal trees have been devised, such as ID3/4/5, CLS, ASSISTANT, and CART.

Advantages and Disadvantages

Among decision support tools, decision trees (and influence diagrams) have several advantages. Decision trees:

- Are simple to understand and interpret. People are able to understand decision tree models after a brief explanation.

- Have value even with little hard data. Important insights can be generated based on experts describing a situation (its alternatives, probabilities, and costs) and their preferences for outcomes.

- Allow the addition of new possible scenarios

- Help determine worst, best and expected values for different scenarios

- Use a white box model. If a given result is provided by a model.

- Can be combined with other decision techniques.

Disadvantages of decision trees:

- For data including categorical variables with different number of levels, information gain in decision trees are biased in favor of those attributes with more levels.

- Calculations can get very complex particularly if many values are uncertain and/or if many outcomes are linked.

Decision trees are a class of learning models that are more robust to noise as well as more powerful as compared to concept learning. Consider the problem of classifying a star based on some astronomical measurements. It can naturally be represented by the following set of decisions on each measurement arranged in a tree like fashion.

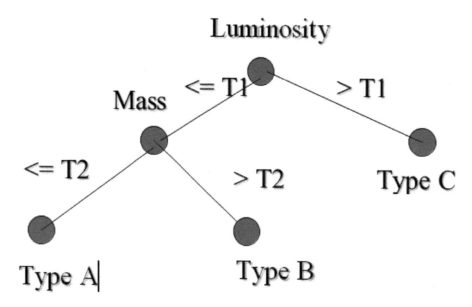

1. A decision-tree learning algorithm approximates a target concept using a tree representation, where each internal node corresponds to an attribute, and every terminal node corresponds to a class.

2. There are two types of nodes:

   - Internal node.- Splits into different branches according to the different values the corresponding attribute can take. Example: luminosity <= T1 or luminosity > T1.

   - Terminal Node.- Decides the class assigned to the example.

Classifying Examples using Decision Tree

To classify an example X we start at the root of the tree, and check the value of that attribute on X. We follow the branch corresponding to that value and jump to the next node. We continue until we reach a terminal node and take that class as our best prediction.

$$X = (\text{Luminosity} <= T1, \text{Mass} > T2)$$

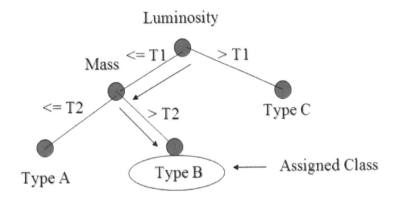

Decision trees adopt a DNF (Disjunctive Normal Form) representation. For a fixed class, every branch from the root of the tree to a terminal node with that class is a conjunction of attribute values; different branches ending in that class form a disjunction.

In the following example, the rules for class A are: (~X1 & ~x2) OR (X1 & ~x3)

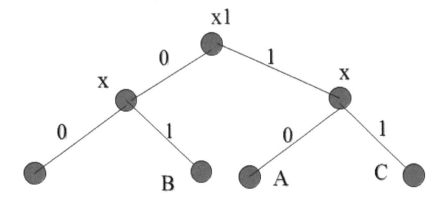

Decision Tree Construction

There are different ways to construct trees from data. We will concentrate on the top- down, greedy search approach:

Basic idea:

1. Choose the best attribute a* to place at the root of the tree.

2. Separate training set D into subsets {D1, D2, .., Dk} where each subset Di contains examples having the same value for a*

3. Recursively apply the algorithm on each new subset until examples have the same class or there are few of them.

Illustration:

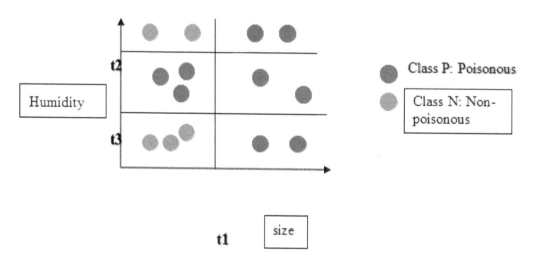

Attributes: size and humidity. Size has two values: >t1 or <= t1

Humidity has three values: >t2, (>t3 and <=t2), <= t3

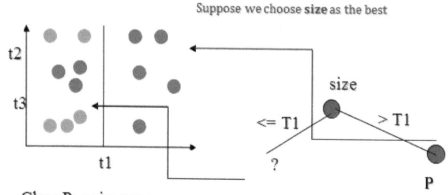

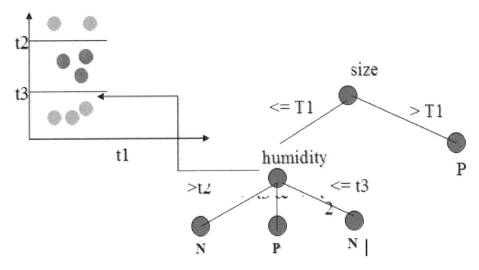

Suppose we choose **humidity** as the next best

Step:

- Create a root for the tree

- If all examples are of the same class or the number of examples is below a threshold return that class

- If no attributes available return majority class

- Let a* be the best attribute

- For each possible value v of a*

  - Add a branch below a* labeled "a = v"

  - Let Sv be the subsets of example where attribute a*=v

  - Recursively apply the algorithm to Sv

Splitting Functions

What attribute is the best to split the data? Let us remember some definitions from information theory.

A measure of uncertainty or entropy that is associated to a random variable X is defined as

$$H(X) = - \Sigma \, pi \log pi$$

where the logarithm is in base 2.

This is the "average amount of information or entropy of a finite complete probability scheme".

We will use a entropy based splitting function.

Consider the previous example:

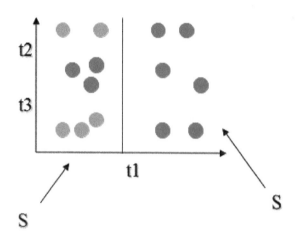

Size divides the sample in two.

S1 = { 6P, 0NP}
S2 = { 3P, 5NP}

H(S1) = 0

H(S2) = -(3/8)log2(3/8)
            -(5/8)log2(5/8)

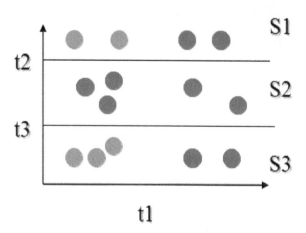

humidity divides the sample in three.

S1 = { 2P, 2NP}
S2 = { 5P, 0NP}
S3 = { 2P, 3NP}

H(S1) = 1
H(S2) = 0
H(S3) = -(2/5)log2(2/5)
            -(3/5)log2(3/5)

Let us define *information gain* as follows:

Information gain IG over attribute A: IG (A) IG(A) = H(S) - Σv (Sv/S) H (Sv)

H(S) is the entropy of all examples. H(Sv) is the entropy of one subsample after partitioning S based on all possible values of attribute A.

Consider the previous example:

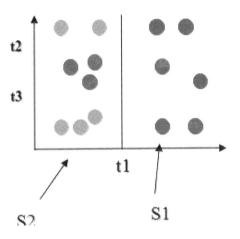

We have,

H(S1) = 0
H(S2) = -(3/8)log2(3/8)
        -(5/8)log2(5/8)

H(S) = -(9/14)log2(9/14)
        -(5/14)log2(5/14)

|S1|/|S| = 6/14
|S2|/|S| = 8/14

The principle for decision tree construction may be stated as follows:

Order the splits (attribute and value of the attribute) in decreasing order of information gain.

Decision Tree Learning

Decision tree learning uses a decision tree as a predictive model observations about an item (represented in the branches) to conclusions about the item's target value (represented in the leaves). It is one of the predictive modelling approaches used in statistics, data mining and machine learning. Tree models where the target variable can take a finite set of values are called classification trees; in these tree structures, leaves represent class labels and branches represent conjunctions of features that lead to those class labels. Decision trees where the target variable can take continuous values (typically real numbers) are called regression trees.

In decision analysis, a decision tree can be used to visually and explicitly represent decisions and decision making. In data mining, a decision tree describes data (but the resulting classification tree can be an input for decision making). This page deals with decision trees in data mining.

General

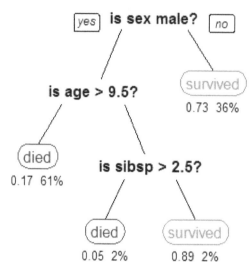

A tree showing survival of passengers on the Titanic ("sibsp" is the number of spouses or siblings aboard). The figures under the leaves show the probability of outcome and the percentage of observations in the leaf.

Decision tree learning is a method commonly used in data mining. The goal is to create a model that predicts the value of a target variable based on several input variables. An example is shown in the diagram. Each interior node corresponds to one of the input variables; there are edges to children for each of the possible values of that input variable. Each leaf represents a value of the target variable given the values of the input variables represented by the path from the root to the leaf.

A decision tree is a simple representation for classifying examples. For this section, assume that all of the input features have finite discrete domains, and there is a single target feature called the classification. Each element of the domain of the classification is called a class. A decision tree or a classification tree is a tree in which each internal (non-leaf) node is labeled with an input feature. The arcs coming from a node labeled with an input feature are labeled with each of the possible values of the target or output feature or the arc leads to a subordinate decision node on a different input feature. Each leaf of the tree is labeled with a class or a probability distribution over the classes.

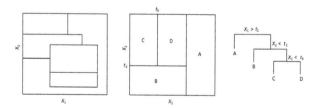

Left: A partitioned two-dimensional feature space. These partitions could not have resulted from recursive binary splitting. Middle: A partitioned two-dimensional feature space with partitions that did result from recursive binary splitting. Right: A tree corresponding to the partitioned feature space in the middle. Notice the convention that when the expression at the split is true, the tree follows the left branch. When the expression is false, the right branch is followed.

A tree can be "learned" by splitting the source set into subsets based on an attribute value test. This process is repeated on each derived subset in a recursive manner called recursive partitioning. The examples illustrated in the figure for spaces that have and have not been partitioned using recursive partitioning, or recursive binary splitting. The recursion is completed when the subset at a node has all the same value of the target variable, or when splitting no longer adds value to the predictions. This process of *top-down induction of decision trees* (TDIDT) is an example of a greedy algorithm, and it is by far the most common strategy for learning decision trees from data.

In data mining, decision trees can be described also as the combination of mathematical and computational techniques to aid the description, categorization and generalization of a given set of data.

Data comes in records of the form:

$$(\mathbf{x}, Y) = (x_1, x_2, x_3, ..., x_k, Y)$$

The dependent variable, Y, is the target variable that we are trying to understand, classify or generalize. The vector x is composed of the input variables, x_1, x_2, x_3 etc., that are used for that task.

Types

Decision trees used in data mining are of two main types:

- Classification tree analysis is when the predicted outcome is the class to which the data belongs.

- Regression tree analysis is when the predicted outcome can be considered a real number (e.g. the price of a house, or a patient's length of stay in a hospital).

The term Classification And Regression Tree (CART) analysis is an umbrella term used to refer to both of the above procedures, first introduced by Breiman et al. Trees used for regression and trees used for classification have some similarities - but also some differences, such as the procedure used to determine where to split.

Some techniques, often called *ensemble* methods, construct more than one decision tree:

- Boosted Trees Incrementally building an ensemble by training each new instance to emphasize the training instances previously mis-modeled. A typical example is AdaBoost. Theses can be used for regression-type and classification-type problems.

- Bagging decision trees, an early ensemble method, builds multiple decision trees by repeatedly resampling training data with replacement, and voting the trees for a consensus prediction.

- A Random Forest classifier is a specific type of Bootstrap aggregating

- Rotation forest - in which every decision tree is trained by first applying principal component analysis (PCA) on a random subset of the input features.

A special case of a decision tree is a Decision list, which is a one-sided decision tree, so that every internal node has exactly 1 leaf node and exactly 1 internal node as a child (except for the bottom-most node, whose only child is a single leaf node). While less expressive, decision lists are arguably easier to understand than general decision trees due to their added sparsity, permit non-greedy learning methods and monotonic constraints to be imposed.

Decision tree learning is the construction of a decision tree from class-labeled training tuples. A decision tree is a flow-chart-like structure, where each internal (non-leaf) node denotes a test on an attribute, each branch represents the outcome of a test, and each leaf (or terminal) node holds a class label. The topmost node in a tree is the root node.

There are many specific decision-tree algorithms. Notable ones include:

- ID3 (Iterative Dichotomiser 3)

- C4.5 (successor of ID3)

- CART (Classification And Regression Tree)

- CHAID (CHi-squared Automatic Interaction Detector). Performs multi-level splits when computing classification trees.

- MARS: extends decision trees to handle numerical data better.

- Conditional Inference Trees. Statistics-based approach that uses non-parametric tests as splitting criteria, corrected for multiple testing to avoid overfitting. This approach results in unbiased predictor selection and does not require pruning.

ID3 and CART were invented independently at around the same time (between 1970 and 1980), yet follow a similar approach for learning decision tree from training tuples.

Metrics

Algorithms for constructing decision trees usually work top-down, by choosing a variable at each step that best splits the set of items. Different algorithms use different metrics for measuring "best". These generally measure the homogeneity of the target variable within the subsets. Some examples are given below. These metrics are applied to each candidate subset, and the resulting values are combined (e.g., averaged) to provide a measure of the quality of the split.

Gini Impurity

Used by the CART (classification and regression tree) algorithm, Gini impurity is a measure of how often a randomly chosen element from the set would be incorrectly labeled if it was randomly labeled according to the distribution of labels in the subset. Gini impurity can be computed by summing the probability p_i of an item with label i being chosen times the probability $1 - p_i$ of a mistake in categorizing that item. It reaches its minimum (zero) when all cases in the node fall into a single target category.

To compute Gini impurity for a set of items with J classes, suppose $i \in \{1, 2, ..., J\}$, and let p_i be the fraction of items labeled with class i in the set.

$$I_G(p) = \sum_{i=1}^{J} p_i(1-p_i) = \sum_{i=1}^{J}(p_i - p_i^2) = \sum_{i=1}^{J} p_i - \sum_{i=1}^{J} p_i^2 = 1 - \sum_{i=1}^{J} p_i^2 = \sum_{i \neq k} p_i p_k$$

Information Gain

Used by the ID3, C4.5 and C5.0 tree-generation algorithms. Information gain is based on the concept of entropy from information theory.

Entropy is defined as below

$$H(T) = I_E(p_1, p_2, ..., p_n) = -\sum_{i=1}^{J} p_i \log_2 p_i$$

where $p_1, p_2, \cdots$ are fractions that add up to 1 and represent the percentage of each class present in the child node that results from a split in the tree.

Information Gain = Entropy(parent) - Weighted Sum of Entropy(Children)

$$IG(T, a) = H(T) - H(T \mid a)$$

Information gain is used to decide which feature to split on at each step in building the tree. Simplicity is best, so we want to keep our tree small. To do so, at each step we should choose the split that results in the purest daughter nodes. A commonly used measure of purity is called information which is measured in bits, not to be confused with the unit of computer memory. For each node of the tree, the information value "represents the expected amount of information that would be needed to specify whether a new instance should be classified yes or no, given that the example reached that node".

Consider an example data set with four attributes: outlook (sunny, overcast, rainy), temperature (hot, mild, cool), humidity (high, normal), and windy (true, false), with a binary (yes or no) target variable, play, and 14 data points. To construct a decision tree on this data, we need to compare the information gain of each of four trees, each split on one of the four features. The split with the highest information gain will be taken as the first split and the process will continue until all children nodes are pure, or until the information gain is 0.

The split using the feature windy results in two children nodes, one for a windy value of true and one for a windy value of false. In this data set, there are six data points with a true windy value, three of which have a play value of yes and three with a play value of no. The eight remaining data points with a windy value of false contain two no's and six yes's. The information of the windy=true node is calculated using the entropy equation above. Since there is an equal number of yes's and no's in this node, we have

$$I_E([3,3]) = -(3/6)\log_2(3/6) - (3/6)\log_2(3/6) = -(1/2)\log_2(1/2) - (1/2)\log_2(1/2) = 1$$

For the node where windy=false there were eight data points, six yes's and two no's. Thus we have

$$I_E([6,2]) = -(6/8)\log_2(6/8) - (2/8)\log_2(2/8) = -(3/4)\log_2(3/4) - (1/4)\log_2(1/4) = 0.8112781$$

To find the information of the split, we take the weighted average of these two numbers based on how many observations fell into which node.

$$I_E([3,3],[6,2]) = I_E(windyornot) = (6/14)(1) + (8/14)(0.8112781) = 0.8921589$$

To find the information gain of the split using windy, we must first calculate the information in the data before the split. The original data contained nine yes's and five no's.

$$I_E([9,5]) = -(9/14)\log_2(9/14) - (5/14)\log_2(5/14) = 0.940286$$

Now we can calculate the information gain achieved by splitting on the windy feature.

$$IG(windy) = I_E([9,5]) - I_E([3,3],[6,2]) = 0.940286 - 0.8921589 = 0.0481271$$

To build the tree, the information gain of each possible first split would need to be calculated. The best first split is the one that provides the most information gain. This process is repeated for each impure node until the tree is complete. This example is adapted from the example appearing in Witten et al.

Variance Reduction

Introduced in CART, variance reduction is often employed in cases where the target variable is continuous (regression tree), meaning that use of many other metrics would first require discretization before being applied. The variance reduction of a node N is defined as the total reduction of the variance of the target variable x due to the split at this node:

$$I_V(N) = \frac{1}{|S|^2}\sum_{i\in S}\sum_{j\in S}\frac{1}{2}(x_i - x_j)^2 - \left(\frac{1}{|S_t|^2}\sum_{i\in S_t}\sum_{j\in S_t}\frac{1}{2}(x_i - x_j)^2 + \frac{1}{|S_f|^2}\sum_{i\in S_f}\sum_{j\in S_f}\frac{1}{2}(x_i - x_j)^2\right)$$

where S, S_t, and S_f are the set of presplit sample indices, set of sample indices for which the split test is true, and set of sample indices for which the split test is false, respectively. Each of the above summands are indeed variance estimates, though, written in a form without directly referring to the mean.

Decision Tree Advantages

Amongst other data mining methods, decision trees have various advantages:

- Simple to understand and interpret. People are able to understand decision tree models after a brief explanation. Trees can also be displayed graphically in a way that is easy for non-experts to interpret.

- Able to handle both numerical and categorical data. Other techniques are usually specialised in analysing datasets that have only one type of variable. (For example, relation rules can be used only with nominal variables while neural networks can be used only with numerical variables or categoricals converted to 0-1 values.)

- Requires little data preparation. Other techniques often require data normalization. Since trees can handle qualitative predictors, there is no need to create dummy variables.

- Uses a white box model. If a given situation is observable in a model the explanation for the condition is easily explained by boolean logic. By contrast, in a black box model, the explanation for the results is typically difficult to understand, for example with an artificial neural network.

- Possible to validate a model using statistical tests. That makes it possible to account for the reliability of the model.

- Non-statistical approach that makes no assumptions of the training data or prediction residuals; e.g., no distributional, independence, or constant variance assumptions

- Performs well with large datasets. Large amounts of data can be analysed using standard computing resources in reasonable time.

- Mirrors human decision making more closely than other approaches. This could be useful when modeling human decisions/behavior.

Limitations

- Trees do not tend to be as accurate as other approaches.

- Trees can be very non-robust. A small change in the training data can result in a big change in the tree, and thus a big change in final predictions.

- The problem of learning an optimal decision tree is known to be NP-complete under several aspects of optimality and even for simple concepts. Consequently, practical decision-tree learning algorithms are based on heuristics such as the greedy algorithm where locally-optimal decisions are made at each node. Such algorithms cannot guarantee to return the globally-optimal decision tree. To reduce the greedy effect of local-optimality some methods such as the dual information distance (DID) tree were proposed.

- Decision-tree learners can create over-complex trees that do not generalize well from the training data. (This is known as overfitting.) Mechanisms such as pruning are necessary to avoid this problem (with the exception of some algorithms such as the Conditional Inference approach, that does not require pruning).

- There are concepts that are hard to learn because decision trees do not express them easily, such as XOR, parity or multiplexer problems. In such cases, the decision tree becomes prohibitively large. Approaches to solve the problem involve either changing the representation of the problem domain (known as propositionalization) or using learning algorithms based on more expressive representations (such as statistical relational learning or inductive logic programming).

- For data including categorical variables with different numbers of levels, information gain in decision trees is biased in favor of those attributes with more levels. However, the issue of biased predictor selection is avoided by the Conditional Inference approach.

Extensions

Decision Graphs

In a decision tree, all paths from the root node to the leaf node proceed by way of conjunction, or *AND*. In a decision graph, it is possible to use disjunctions (ORs) to join two more paths together using Minimum message length (MML). Decision graphs have been further extended to allow for previously unstated new attributes to be learnt dynamically and used at different places within the graph. The more general coding scheme results in better predictive accuracy and log-loss probabilistic scoring. In general, decision graphs infer models with fewer leaves than decision trees.

Alternative Search Methods

Evolutionary algorithms have been used to avoid local optimal decisions and search the decision tree space with little *a priori* bias.

It is also possible for a tree to be sampled using MCMC.

The tree can be searched for in a bottom-up fashion.

References

- Urbanowicz, Ryan J.; Moore, Jason H. (2009-09-22). "Learning Classifier Systems: A Complete Introduction, Review, and Roadmap". Journal of Artificial Evolution and Applications. 2009: 1–25. ISSN 1687-6229. doi:10.1155/2009/736398

- Russell, Stuart; Norvig, Peter (2003) [1995]. Artificial Intelligence: A Modern Approach (2nd ed.). Prentice Hall. ISBN 978-0137903955

- "Gartner's 2016 Hype Cycle for Emerging Technologies Identifies Three Key Trends That Organizations Must Track to Gain Competitive Advantage". Retrieved 2017-04-10

- Mohri, Mehryar; Rostamizadeh, Afshin; Talwalkar, Ameet (2012). Foundations of Machine Learning. USA, Massachusetts: MIT Press. ISBN 9780262018258

- Kamiński, B.; Jakubczyk, M.; Szufel, P. (2017). "A framework for sensitivity analysis of decision trees". Central European Journal of Operations Research. doi:10.1007/s10100-017-0479-6

- Alpaydin, Ethem (2010). Introduction to Machine Learning. London: The MIT Press. ISBN 978-0-262-01243-0. Retrieved 4 February 2017

- Quinlan, J. R. (1987). "Simplifying decision trees". International Journal of Man-Machine Studies. 27 (3): 221. doi:10.1016/S0020-7373(87)80053-6

- Yoshua Bengio (2009). Learning Deep Architectures for AI. Now Publishers Inc. pp. 1–3. ISBN 978-1-60198-294-0

- Simonite, Tom. "Microsoft says its racist chatbot illustrates how AI isn't adaptable enough to help most businesses". MIT Technology Review. Retrieved 2017-04-10

- Rokach, Lior; Maimon, O. (2008). Data mining with decision trees: theory and applications. World Scientific Pub Co Inc. ISBN 978-9812771711

- K. Karimi and H.J. Hamilton (2011), "Generation and Interpretation of Temporal Decision Rules", International Journal of Computer Information Systems and Industrial Management Applications, Volume 3

- Witten, Ian; Frank, Eibe; Hall, Mark (2011). Data Mining. Burlington, MA: Morgan Kaufmann. pp. 102–103. ISBN 978-0-12-374856-0

- Cornell University Library. "Breiman : Statistical Modeling: The Two Cultures (with comments and a rejoinder by the author)". Retrieved 8 August 2015

- Gareth, James; Witten, Daniela; Hastie, Trevor; Tibshirani, Robert (2015). An Introduction to Statistical Learning. New York: Springer. p. 315. ISBN 978-1-4614-7137-0

- Hothorn, T.; Hornik, K.; Zeileis, A. (2006). "Unbiased Recursive Partitioning: A Conditional Inference Framework". Journal of Computational and Graphical Statistics. 15 (3): 651–674. JSTOR 27594202. doi:10.1198/106186006X133933

- Horváth, Tamás; Yamamoto, Akihiro, eds. (2003). "Inductive Logic Programming". Lecture Notes in Computer Science. 2835. ISBN 978-3-540-20144-1. doi:10.1007/b13700

- Chipman, Hugh A., Edward I. George, and Robert E. McCulloch. "Bayesian CART model search." Journal of the American Statistical Association 93.443 (1998): 935-948

Natural Language Processing and its Techniques

Natural language processing is a sub-field of computer science that is related with the interaction between computer and human language. Its primary use is to make computers perform tasks when inputs are provided using human language. Linguistic concepts such as syntax, discourse, semantics, etc. are used to evaluate the degree of a task that is to be performed. The topics discussed in the chapter are of great importance to broaden the existing knowledge on natural language processing.

Natural Language Processing

Natural language processing (NLP) is a field of computer science, artificial intelligence and computational linguistics concerned with the interactions between computers and human (natural) languages, and, in particular, concerned with programming computers to fruitfully process large natural language corpora. Challenges in natural language processing frequently involve natural language understanding, natural language generation (frequently from formal, machine-readable logical forms), connecting language and machine perception, managing human-computer dialog systems, or some combination thereof.

History

The history of NLP generally started in the 1950s, although work can be found from earlier periods. In 1950, Alan Turing published an article titled "Computing Machinery and Intelligence" which proposed what is now called the Turing test as a criterion of intelligence.

The Georgetown experiment in 1954 involved fully automatic translation of more than sixty Russian sentences into English. The authors claimed that within three or five years, machine translation would be a solved problem. However, real progress was much slower, and after the ALPAC report in 1966, which found that ten-year-long research had failed to fulfill the expectations, funding for machine translation was dramatically reduced. Little further research in machine translation was conducted until the late 1980s, when the first statistical machine translation systems were developed.

Some notably successful NLP systems developed in the 1960s were SHRDLU, a natural language system working in restricted "blocks worlds" with restricted vocabularies, and ELIZA, a simulation of a Rogerian psychotherapist, written by Joseph Weizenbaum between 1964 and 1966. Using almost no information about human thought or emotion, ELIZA sometimes provided a startlingly human-like interaction. When the "patient" exceeded the very small knowledge base, ELIZA might provide a generic response, for example, responding to "My head hurts" with "Why do you say your head hurts?".

During the 1970s, many programmers began to write "conceptual ontologies", which structured real-world information into computer-understandable data. Examples are MARGIE (Schank, 1975), SAM (Cullingford, 1978), PAM (Wilensky, 1978), TaleSpin (Meehan, 1976), QUALM (Lehnert, 1977), Politics (Carbonell, 1979), and Plot Units (Lehnert 1981). During this time, many chatterbots were written including PARRY, Racter, and Jabberwacky.

Up to the 1980s, most NLP systems were based on complex sets of hand-written rules. Starting in the late 1980s, however, there was a revolution in NLP with the introduction of machine learning algorithms for language processing. This was due to both the steady increase in computational power and the gradual lessening of the dominance of Chomskyan theories of linguistics (e.g. transformational grammar), whose theoretical underpinnings discouraged the sort of corpus linguistics that underlies the machine-learning approach to language processing. Some of the earliest-used machine learning algorithms, such as decision trees, produced systems of hard if-then rules similar to existing hand-written rules. However, part-of-speech tagging introduced the use of hidden Markov models to NLP, and increasingly, research has focused on statistical models, which make soft, probabilistic decisions based on attaching real-valued weights to the features making up the input data. The cache language models upon which many speech recognition systems now rely are examples of such statistical models. Such models are generally more robust when given unfamiliar input, especially input that contains errors (as is very common for real-world data), and produce more reliable results when integrated into a larger system comprising multiple subtasks.

Many of the notable early successes occurred in the field of machine translation, due especially to work at IBM Research, where successively more complicated statistical models were developed. These systems were able to take advantage of existing multilingual textual corpora that had been produced by the Parliament of Canada and the European Union as a result of laws calling for the translation of all governmental proceedings into all official languages of the corresponding systems of government. However, most other systems depended on corpora specifically developed for the tasks implemented by these systems, which was (and often continues to be) a major limitation in the success of these systems. As a result, a great deal of research has gone into methods of more effectively learning from limited amounts of data.

Recent research has increasingly focused on unsupervised and semi-supervised learning algorithms. Such algorithms are able to learn from data that has not been hand-annotated with the desired answers, or using a combination of annotated and non-annotated data. Generally, this task is much more difficult than supervised learning, and typically produces less accurate results for a given amount of input data. However, there is an enormous amount of non-annotated data available (including, among other things, the entire content of the World Wide Web), which can often make up for the inferior results.

In recent years, there has been a flurry of results showing deep learning techniques achieving state-of-the-art results in many natural language tasks, for example in language modeling, parsing, and many others.

Statistical Natural Language Processing

Since the so-called "statistical revolution" in the late 1980s and mid 1990s, much Natural Language Processing research has relied heavily on machine learning.

Formerly, many language-processing tasks typically involved the direct hand coding of rules, which is not in general robust to natural language variation. The machine-learning paradigm calls instead for using statistical inference to automatically learn such rules through the analysis of large *corpora* of typical real-world examples (a *corpus* (plural, "corpora") is a set of documents, possibly with human or computer annotations).

Many different classes of machine learning algorithms have been applied to NLP tasks. These algorithms take as input a large set of "features" that are generated from the input data. Some of the earliest-used algorithms, such as decision trees, produced systems of hard if-then rules similar to the systems of hand-written rules that were then common. Increasingly, however, research has focused on statistical models, which make soft, probabilistic decisions based on attaching real-valued weights to each input feature. Such models have the advantage that they can express the relative certainty of many different possible answers rather than only one, producing more reliable results when such a model is included as a component of a larger system.

Systems based on machine-learning algorithms have many advantages over hand-produced rules:

- The learning procedures used during machine learning automatically focus on the most common cases, whereas when writing rules by hand it is often not at all obvious where the effort should be directed.

- Automatic learning procedures can make use of statistical inference algorithms to produce models that are robust to unfamiliar input (e.g. containing words or structures that have not been seen before) and to erroneous input (e.g. with misspelled words or words accidentally omitted). Generally, handling such input gracefully with hand-written rules—or more generally, creating systems of hand-written rules that make soft decisions—is extremely difficult, error-prone and time-consuming.

- Systems based on automatically learning the rules can be made more accurate simply by supplying more input data. However, systems based on hand-written rules can only be made more accurate by increasing the complexity of the rules, which is a much more difficult task. In particular, there is a limit to the complexity of systems based on hand-crafted rules, beyond which the systems become more and more unmanageable. However, creating more data to input to machine-learning systems simply requires a corresponding increase in the number of man-hours worked, generally without significant increases in the complexity of the annotation process.

Major Evaluations and Tasks

The following is a list of some of the most commonly researched tasks in NLP. Note that some of these tasks have direct real-world applications, while others more commonly serve as subtasks that are used to aid in solving larger tasks.

Though NLP tasks are obviously very closely intertwined, they are frequently, for convenience, subdivided into categories. A coarse division is given below.

Syntax

Lemmatization

Morphological segmentation

> Separate words into individual morphemes and identify the class of the morphemes. The difficulty of this task depends greatly on the complexity of the morphology (i.e. the structure of words) of the language being considered. English has fairly simple morphology, especially inflectional morphology, and thus it is often possible to ignore this task entirely and simply model all possible forms of a word (e.g. "open, opens, opened, opening") as separate words. In languages such as Turkish or Meitei, a highly agglutinated Indian language, however, such an approach is not possible, as each dictionary entry has thousands of possible word forms.

Part-of-speech tagging

> Given a sentence, determine the part of speech for each word. Many words, especially common ones, can serve as multiple parts of speech. For example, "book" can be a noun ("the book on the table") or verb ("to book a flight"); "set" can be a noun, verb or adjective; and "out" can be any of at least five different parts of speech. Some languages have more such ambiguity than others. Languages with little inflectional morphology, such as English are particularly prone to such ambiguity. Chinese is prone to such ambiguity because it is a tonal language during verbalization. Such inflection is not readily conveyed via the entities employed within the orthography to convey intended meaning.

Parsing

> Determine the parse tree (grammatical analysis) of a given sentence. The grammar for natural languages is ambiguous and typical sentences have multiple possible analyses. In fact, perhaps surprisingly, for a typical sentence there may be thousands of potential parses (most of which will seem completely nonsensical to a human).

Sentence breaking (also known as sentence boundary disambiguation)

> Given a chunk of text, find the sentence boundaries. Sentence boundaries are often marked by periods or other punctuation marks, but these same characters can serve other purposes (e.g. marking abbreviations).

Stemming

Word segmentation

> Separate a chunk of continuous text into separate words. For a language like English, this is fairly trivial, since words are usually separated by spaces. However, some written languages like Chinese, Japanese and Thai do not mark word boundaries in such a fashion, and in those languages text segmentation is a significant task requiring knowledge of the vocabulary and morphology of words in the language.

Semantics

Lexical semantics

What is the computational meaning of individual words in context?

Machine translation

Automatically translate text from one human language to another. This is one of the most difficult problems, and is a member of a class of problems colloquially termed "AI-complete", i.e. requiring all of the different types of knowledge that humans possess (grammar, semantics, facts about the real world, etc.) in order to solve properly.

Named entity recognition (NER)

Given a stream of text, determine which items in the text map to proper names, such as people or places, and what the type of each such name is (e.g. person, location, organization). Note that, although capitalization can aid in recognizing named entities in languages such as English, this information cannot aid in determining the type of named entity, and in any case is often inaccurate or insufficient. For example, the first word of a sentence is also capitalized, and named entities often span several words, only some of which are capitalized. Furthermore, many other languages in non-Western scripts (e.g. Chinese or Arabic) do not have any capitalization at all, and even languages with capitalization may not consistently use it to distinguish names. For example, German capitalizes all nouns, regardless of whether they refer to names, and French and Spanish do not capitalize names that serve as adjectives.

Natural language generation

Convert information from computer databases or semantic intents into readable human language.

Natural language understanding

Convert chunks of text into more formal representations such as first-order logic structures that are easier for computer programs to manipulate. Natural language understanding involves the identification of the intended semantic from the multiple possible semantics which can be derived from a natural language expression which usually takes the form of organized notations of natural languages concepts. Introduction and creation of language metamodel and ontology are efficient however empirical solutions. An explicit formalization of natural languages semantics without confusions with implicit assumptions such as closed-world assumption (CWA) vs. open-world assumption, or subjective Yes/No vs. objective True/False is expected for the construction of a basis of semantics formalization.

Optical character recognition (OCR)

Given an image representing printed text, determine the corresponding text.

Question answering

Given a human-language question, determine its answer. Typical questions have a specific right answer (such as "What is the capital of Canada?"), but sometimes open-ended questions are also considered (such as "What is the meaning of life?"). Recent works have looked at even more complex questions.

Recognizing Textual entailment

Given two text fragments, determine if one being true entails the other, entails the other's negation, or allows the other to be either true or false.

Relationship extraction

Given a chunk of text, identify the relationships among named entities (e.g. who is married to whom).

Sentiment analysis

Extract subjective information usually from a set of documents, often using online reviews to determine "polarity" about specific objects. It is especially useful for identifying trends of public opinion in the social media, for the purpose of marketing.

Topic segmentation and recognition

Given a chunk of text, separate it into segments each of which is devoted to a topic, and identify the topic of the segment.

Word sense disambiguation

Many words have more than one meaning; we have to select the meaning which makes the most sense in context. For this problem, we are typically given a list of words and associated word senses, e.g. from a dictionary or from an online resource such as WordNet.

Discourse

Automatic summarization

Produce a readable summary of a chunk of text. Often used to provide summaries of text of a known type, such as articles in the financial section of a newspaper.

Coreference resolution

Given a sentence or larger chunk of text, determine which words ("mentions") refer to the same objects ("entities"). Anaphora resolution is a specific example of this task, and is specifically concerned with matching up pronouns with the nouns or names to which they refer. The more general task of coreference resolution also includes identifying so-called "bridging relationships" involving referring expressions. For example, in a sentence such as "He entered John's house through the front door", "the front door" is a referring expression and the bridging relationship to be identified is the fact that the door being referred to is the front door of John's house (rather than of some other structure that might also be referred to).

Discourse analysis

This rubric includes a number of related tasks. One task is identifying the discourse structure of connected text, i.e. the nature of the discourse relationships between sentences (e.g. elaboration, explanation, contrast). Another possible task is recognizing and classifying the speech acts in a chunk of text (e.g. yes-no question, content question, statement, assertion, etc.).

Speech

Speech recognition

Given a sound clip of a person or people speaking, determine the textual representation of the speech. This is the opposite of text to speech and is one of the extremely difficult problems colloquially termed "AI-complete". In natural speech there are hardly any pauses between successive words, and thus speech segmentation is a necessary subtask of speech recognition. Note also that in most spoken languages, the sounds representing successive letters blend into each other in a process termed coarticulation, so the conversion of the analog signal to discrete characters can be a very difficult process.

Speech segmentation

Given a sound clip of a person or people speaking, separate it into words. A subtask of speech recognition and typically grouped with it.

Natural Language Understanding

The steps in natural language understanding are as follows:

Words

↓

Morphological Analysis

Morphologically analyzed words   *(another step: POS tagging)*

↓

Syntactic Analysis

Syntactic Structure

↓

Semantic Analysis

Context-independent meaning representation

↓

Discourse Processing

Final meaning representation

↓

Natural language understanding (NLU) is a subtopic of natural language processing in artificial intelligence that deals with machine reading comprehension. NLU is considered an AI-hard problem.

The process of disassembling and parsing input is more complex than the reverse process of assembling output in natural language generation because of the occurrence of unknown and unexpected features in the input and the need to determine the appropriate syntactic and semantic schemes to apply to it, factors which are pre-determined when outputting language.

There is considerable commercial interest in the field because of its application to news-gathering, text categorization, voice-activation, archiving, and large-scale content-analysis.

History

The program STUDENT, written in 1964 by Daniel Bobrow for his PhD dissertation at MIT is one of the earliest known attempts at natural language understanding by a computer. Eight years after John McCarthy coined the term artificial intelligence, Bobrow's dissertation (titled *Natural Language Input for a Computer Problem Solving System*) showed how a computer can understand simple natural language input to solve algebra word problems.

A year later, in 1965, Joseph Weizenbaum at MIT wrote ELIZA, an interactive program that carried on a dialogue in English on any topic, the most popular being psychotherapy. ELIZA worked by simple parsing and substitution of key words into canned phrases and Weizenbaum sidestepped the problem of giving the program a database of real-world knowledge or a rich lexicon. Yet ELIZA gained surprising popularity as a toy project and can be seen as a very early precursor to current commercial systems such as those used by Ask.com.

In 1969 Roger Schank at Stanford University introduced the conceptual dependency theory for natural language understanding. This model, partially influenced by the work of Sydney Lamb, was extensively used by Schank's students at Yale University, such as Robert Wilensky, Wendy Lehnert, and Janet Kolodner.

In 1970, William A. Woods introduced the augmented transition network (ATN) to represent natural language input. Instead of *phrase structure rules* ATNs used an equivalent set of finite state automata that were called recursively. ATNs and their more general format called "generalized ATNs" continued to be used for a number of years.

In 1971 Terry Winograd finished writing SHRDLU for his PhD thesis at MIT. SHRDLU could understand simple English sentences in a restricted world of children's blocks to direct a robotic arm to move items. The successful demonstration of SHRDLU provided significant momentum for continued research in the field. Winograd continued to be a major influence in the field with the publication of his book *Language as a Cognitive Process*. At Stanford, Winograd would later be the adviser for Larry Page, who co-founded Google.

In the 1970s and 1980s the natural language processing group at SRI International continued research and development in the field. A number of commercial efforts based on the research were undertaken, *e.g.*, in 1982 Gary Hendrix formed Symantec Corporation originally as a company for developing a natural language interface for database queries on personal computers. However, with the advent of mouse driven, graphic user interfaces Symantec changed direction. A number of other commercial efforts were started around the same time, *e.g.*, Larry R. Harris at the Artificial Intelligence Corporation and Roger Schank and his students at Cognitive Systems corp. In 1983, Michael Dyer developed the BORIS system at Yale which bore similarities to the work of Roger Schank and W. G. Lehnart.

The third millennium saw the introduction of systems using machine learning for text classification, such as the IBM Watson. However, this is not NLU. According to John Searle, Watson did not even understand the questions.

John Ball, cognitive scientist and inventor of Patom Theory supports this assessment. NLP has made inroads for applications to support human productivity in service and ecommerce but this has largely been made possible by narrowing the scope of the application. There are thousands of ways to request something in a human language which still defies conventional NLP. "To have a meaningful conversation with machines is only possible when we match every word to the correct meaning based on the meanings of the other words in the sentence – just like a 3-year-old does without guesswork" Patom Theory

Scope and Context

The umbrella term "natural language understanding" can be applied to a diverse set of computer applications, ranging from small, relatively simple tasks such as short commands issued to robots, to highly complex endeavors such as the full comprehension of newspaper articles or poetry passages. Many real world applications fall between the two extremes, for instance text classification for the automatic analysis of emails and their routing to a suitable department in a corporation does not require in depth understanding of the text, but is far more complex than the management of simple queries to database tables with fixed schemata.

Throughout the years various attempts at processing natural language or *English-like* sentences presented to computers have taken place at varying degrees of complexity. Some attempts have not resulted in systems with deep understanding, but have helped overall system usability. For example, Wayne Ratliff originally developed the *Vulcan* program with an English-like syntax to mimic the English speaking computer in Star Trek. Vulcan later became the dBase system whose easy-to-use syntax effectively launched the personal computer database industry. Systems with an easy to use or *English like* syntax are, however, quite distinct from systems that use a rich lexicon and include an internal representation (often as first order logic) of the semantics of natural language sentences.

Hence the breadth and depth of "understanding" aimed at by a system determine both the complexity of the system (and the implied challenges) and the types of applications it can deal with. The "breadth" of a system is measured by the sizes of its vocabulary and grammar. The "depth" is measured by the degree to which its understanding approximates that of a fluent native speaker. At the narrowest and shallowest, *English-like* command interpreters require minimal complexity, but have a small range of applications. Narrow but deep systems explore and model mechanisms of understanding, but they still have limited application. Systems that attempt to understand the contents of a document such as a news release beyond simple keyword matching and to judge its suitability for a user are broader and require significant complexity, but they are still somewhat shallow. Systems that are both very broad and very deep are beyond the current state of the art.

Components and Architecture

Regardless of the approach used, most natural language understanding systems share some common components. The system needs a lexicon of the language and a parser and grammar rules to break sentences into an internal representation. The construction of a rich lexicon with a suitable ontology requires significant effort, *e.g.*, the Wordnet lexicon required many person-years of effort.

The system also needs a *semantic theory* to guide the comprehension. The interpretation capabilities of a language understanding system depend on the semantic theory it uses. Competing semantic theories of language have specific trade offs in their suitability as the basis of computer-automated semantic interpretation. These range from *naive semantics* or *stochastic semantic analysis* to the use of *pragmatics* to derive meaning from context.

Advanced applications of natural language understanding also attempt to incorporate logical inference within their framework. This is generally achieved by mapping the derived meaning into a set of assertions in predicate logic, then using logical deduction to arrive at conclusions. Therefore, systems based on functional languages such as Lisp need to include a subsystem to represent logical assertions, while logic-oriented systems such as those using the language Prolog generally rely on an extension of the built-in logical representation framework.

The management of context in natural language understanding can present special challenges. A large variety of examples and counter examples have resulted in multiple approaches to the formal modeling of context, each with specific strengths and weaknesses.

Parsing

Parsing syntax analysis or syntactic analysis is the process of analysing a string of symbols, either in natural language or in computer languages, conforming to the rules of a formal grammar. The term *parsing* comes from Latin *pars* (*orationis*), meaning part (of speech).

The term has slightly different meanings in different branches of linguistics and computer science. Traditional sentence parsing is often performed as a method of understanding the exact meaning of a sentence or word, sometimes with the aid of devices such as sentence diagrams. It usually emphasizes the importance of grammatical divisions such as subject and predicate.

Within computational linguistics the term is used to refer to the formal analysis by a computer of a sentence or other string of words into its constituents, resulting in a parse tree showing their syntactic relation to each other, which may also contain semantic and other information.

The term is also used in psycholinguistics when describing language comprehension. In this context, parsing refers to the way that human beings analyze a sentence or phrase (in spoken language or text) "in terms of grammatical constituents, identifying the parts of speech, syntactic relations, etc." This term is especially common when discussing what linguistic cues help speakers to interpret garden-path sentences.

Within computer science, the term is used in the analysis of computer languages, referring to the syntactic analysis of the input code into its component parts in order to facilitate the writing of compilers and interpreters. The term may also be used to describe a split or separation.

Human Languages

Traditional Methods

The traditional grammatical exercise of parsing, sometimes known as *clause analysis*, involves breaking down a text into its component parts of speech with an explanation of the form, function, and syntactic relationship of each part. This is determined in large part from study of the language's conjugations and declensions, which can be quite intricate for heavily inflected languages. To parse a phrase such as 'man bites dog' involves noting that the singular noun 'man' is the subject of the sentence, the verb 'bites' is the third person singular of the present tense of the verb 'to bite', and the singular noun 'dog' is the object of the sentence. Techniques such as sentence diagrams are sometimes used to indicate relation between elements in the sentence.

Parsing was formerly central to the teaching of grammar throughout the English-speaking world, and widely regarded as basic to the use and understanding of written language. However, the general teaching of such techniques is no longer current.

Computational Methods

Normally parsing is defined as separation. To separate the sentence into grammatical meaning or words, phrase, numbers. In some machine translation and natural language processing systems, written texts in human languages are parsed by computer programs. Human sentences are not easily parsed by programs, as there is substantial ambiguity in the structure of human language, whose usage is to convey meaning (or semantics) amongst a potentially unlimited range of possibilities but only some of which are germane to the particular case. So an utterance "Man bites dog" versus "Dog bites man" is definite on one detail but in another language might appear as "Man dog bites" with a reliance on the larger context to distinguish between those two possibilities, if indeed that difference was of concern. It is difficult to prepare formal rules to describe informal behaviour even though it is clear that some rules are being followed.

In order to parse natural language data, researchers must first agree on the grammar to be used. The choice of syntax is affected by both linguistic and computational concerns; for instance some parsing systems use lexical functional grammar, but in general, parsing for grammars of this type is known to be NP-complete. Head-driven phrase structure grammar is another linguistic formalism which has been popular in the parsing community, but other research efforts have focused on less complex formalisms such as the one used in the Penn Treebank. Shallow parsing aims to find only the boundaries of major constituents such as noun phrases. Another popular strategy for avoiding linguistic controversy is dependency grammar parsing.

Most modern parsers are at least partly statistical; that is, they rely on a corpus of training data which has already been annotated (parsed by hand). This approach allows the system to gather information about the frequency with which various constructions occur in specific contexts. Approaches which have been used include straightforward PCFGs (probabilistic context-free grammars), maximum entropy, and neural nets. Most of the more successful systems use *lexical* statistics (that is, they consider the identities of the words involved, as well as their part of speech). However such systems are vulnerable to overfitting and require some kind of smoothing to be effective.

Parsing algorithms for natural language cannot rely on the grammar having 'nice' properties as with manually designed grammars for programming languages. As mentioned earlier some grammar formalisms are very difficult to parse computationally; in general, even if the desired structure is not context-free, some kind of context-free approximation to the grammar is used to perform a first pass. Algorithms which use context-free grammars often rely on some variant of the CYK algorithm, usually with some heuristic to prune away unlikely analyses to save time. However some systems trade speed for accuracy using, e.g., linear-time versions of the shift-reduce algorithm. A somewhat recent development has been parse reranking in which the parser proposes some large number of analyses, and a more complex system selects the best option.

Psycholinguistics

In psycholinguistics, parsing involves not just the assignment of words to categories, but the evaluation of the meaning of a sentence according to the rules of syntax drawn by inferences made from each word in the sentence. This normally occurs as words are being heard or read. Consequently, psycholinguistic models of parsing are of necessity *incremental*, meaning that they build up an interpretation as the sentence is being processed, which is normally expressed in terms of a partial syntactic structure. Creation of initially wrong structures occurs when interpreting garden path sentences.

Computer Languages

Parser

A parser is a software component that takes input data (frequently text) and builds a data structure – often some kind of parse tree, abstract syntax tree or other hierarchical structure – giving a structural representation of the input, checking for correct syntax in the process. The parsing may be preceded or followed by other steps, or these may be combined into a single step. The parser is often preceded by a separate lexical analyser, which creates tokens from the sequence of input characters; alternatively, these can be combined in scannerless parsing. Parsers may be programmed by hand or may be automatically or semi-automatically generated by a parser generator. Parsing is complementary to templating, which produces formatted *output*. These may be applied to different domains, but often appear together, such as the scanf/printf pair, or the input (front end parsing) and output (back end code generation) stages of a compiler.

The input to a parser is often text in some computer language, but may also be text in a natural language or less structured textual data, in which case generally only certain parts of the text are extracted, rather than a parse tree being constructed. Parsers range from very simple functions such as scanf, to complex programs such as the frontend of a C++ compiler or the HTML parser of a web browser. An important class of simple parsing is done using regular expressions, in which a group of regular expressions defines a regular language and a regular expression engine automatically generating a parser for that language, allowing pattern matching and extraction of text. In other contexts regular expressions are instead used prior to parsing, as the lexing step whose output is then used by the parser.

The use of parsers varies by input. In the case of data languages, a parser is often found as the file reading facility of a program, such as reading in HTML or XML text; these examples are markup languages. In the case of programming languages, a parser is a component of a compiler or interpreter, which parses the source code of a computer programming language to create some form of internal representation; the parser is a key step in the compiler frontend. Programming languages tend to be specified in terms of a deterministic context-free grammar because fast and efficient parsers can be written for them. For compilers, the parsing itself can be done in one pass or multiple passes.

The implied disadvantages of a one-pass compiler can largely be overcome by adding fix-ups, where provision is made for fix-ups during the forward pass, and the fix-ups are applied backwards when the current program segment has been recognized as having been completed. An example where such a fix-up mechanism would be useful would be a forward GOTO statement, where the target of the GOTO is unknown until the program segment is completed. In this case, the application of the fix-up would be delayed until the target of the GOTO was recognized. Obviously, a backward GOTO does not require a fix-up.

Context-free grammars are limited in the extent to which they can express all of the requirements of a language. Informally, the reason is that the memory of such a language is limited. The grammar cannot remember the presence of a construct over an arbitrarily long input; this is necessary for a language in which, for example, a name must be declared before it may be referenced. More powerful grammars that can express this constraint, however, cannot be parsed efficiently. Thus, it is a common strategy to create a relaxed parser for a context-free grammar which accepts a superset of the desired language constructs (that is, it accepts some invalid constructs); later, the unwanted constructs can be filtered out at the semantic analysis (contextual analysis) step.

For example, in Python the following is syntactically valid code:

```
x = 1

print(x)
```

The following code, however, is syntactically valid in terms of the context-free grammar, yielding a syntax tree with the same structure as the previous, but is syntactically invalid in terms of the context-sensitive grammar, which requires that variables be initialized before use:

```
x = 1

print(y)
```

Rather than being analyzed at the parsing stage, this is caught by checking the *values* in the syntax tree, hence as part of *semantic* analysis: context-sensitive syntax is in practice often more easily analyzed as semantics.

Overview of Process

The first stage is the token generation, or lexical analysis, by which the input character stream is split into meaningful symbols defined by a grammar of regular expressions. For example, a calculator program would look at an input such as "12*(3+4)^2" and split it into the tokens 12, *, (, 3, +, 4,), ^, 2, each of which is a meaningful symbol in the context of an arithmetic expression. The lexer would contain rules to tell it that the characters *, +, ^, (and) mark the start of a new token, so meaningless tokens like "12*" or "(3" will not be generated.

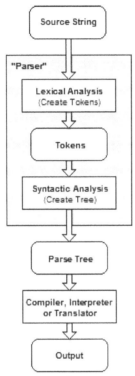

The following example demonstrates the common case of parsing a computer language with two levels of grammar: lexical and syntactic.

The next stage is parsing or syntactic analysis, which is checking that the tokens form an allowable expression. This is usually done with reference to a context-free grammar which recursively defines components that can make up an expression and the order in which they must appear. However, not all rules defining programming languages can be expressed by context-free grammars alone, for example type validity and proper declaration of identifiers. These rules can be formally expressed with attribute grammars.

The final phase is semantic parsing or analysis, which is working out the implications of the expression just validated and taking the appropriate action. In the case of a calculator or interpreter, the action is to evaluate the expression or program, a compiler, on the other hand, would generate some kind of code. Attribute grammars can also be used to define these actions.

Types of Parsers

The *task* of the parser is essentially to determine if and how the input can be derived from the start symbol of the grammar. This can be done in essentially two ways:

- Top-down parsing - Top-down parsing can be viewed as an attempt to find left-most derivations of an input-stream by searching for parse trees using a top-down expansion of the given formal grammar rules. Tokens are consumed from left to right. Inclusive choice is used to accommodate ambiguity by expanding all alternative right-hand-sides of grammar rules.

- Bottom-up parsing - A parser can start with the input and attempt to rewrite it to the start symbol. Intuitively, the parser attempts to locate the most basic elements, then the elements containing these, and so on. LR parsers are examples of bottom-up parsers. Another term used for this type of parser is Shift-Reduce parsing.

LL parsers and recursive-descent parser are examples of top-down parsers which cannot accommodate left recursive production rules. Although it has been believed that simple implementations of top-down parsing cannot accommodate direct and indirect left-recursion and may require exponential time and space complexity while parsing ambiguous context-free grammars, more sophisticated algorithms for top-down parsing have been created by Frost, Hafiz, and Callaghan which accommodate ambiguity and left recursion in polynomial time and which generate polynomial-size representations of the potentially exponential number of parse trees. Their algorithm is able to produce both left-most and right-most derivations of an input with regard to a given context-free grammar.

An important distinction with regard to parsers is whether a parser generates a *leftmost derivation* or a *rightmost derivation*. LL parsers will generate a leftmost derivation and LR parsers will generate a rightmost derivation (although usually in reverse).

Parser Development Software

Some of the well known parser development tools include the following.

- ANTLR

- Bison

- Coco/R

- GOLD

- JavaCC

- Lemon

- Lex

- LuZc

- Parboiled

- Parsec

- Ragel

- Spirit Parser Framework

- Syntax Definition Formalism

- SYNTAX

- XPL

- Yacc

- PackCC

Lookahead

Lookahead establishes the maximum incoming tokens that a parser can use to decide which rule it should use. Lookahead is especially relevant to LL, LR, and LALR parsers, where it is often explicitly indicated by affixing the lookahead to the algorithm name in parentheses, such as LALR(1).

Most programming languages, the primary target of parsers, are carefully defined in such a way that a parser with limited lookahead, typically one, can parse them, because parsers with limited lookahead are often more efficient. One important change to this trend came in 1990 when Terence Parr created ANTLR for his Ph.D. thesis, a parser generator for efficient LL(k) parsers, where k is any fixed value.

Parsers typically have only a few actions after seeing each token. They are shift (add this token to the stack for later reduction), reduce (pop tokens from the stack and form a syntactic construct), end, error (no known rule applies) or conflict (does not know whether to shift or reduce).

Lookahead has two advantages.

- It helps the parser take the correct action in case of conflicts. For example, parsing the if statement in the case of an else clause.

- It eliminates many duplicate states and eases the burden of an extra stack. A C language non-lookahead parser will have around 10,000 states. A lookahead parser will have around 300 states.

Example: Parsing the Expression 1 + 2 * 3

Set of expression parsing rules (called grammar) is as follows,

Rule1: $E \rightarrow E + E$ Expression is the sum of two expressions.

Rule2: $E \rightarrow E * E$ Expression is the product of two expressions.

Rule3: $E \rightarrow$ number   Expression is a simple number

Rule4:   + has less precedence than *

Most programming languages (except for a few such as APL and Smalltalk) and algebraic formulas give higher precedence to multiplication than addition, in which case the correct interpretation of the example above is (1 + (2*3)). Note that Rule4 above is a semantic rule. It is possible to rewrite the grammar to incorporate this into the syntax. However, not all such rules can be translated into syntax.

Simple non-lookahead parser actions

Initially Input = [1,+,2,*,3]

1. Shift "1" onto stack from input (in anticipation of rule3). Input = [+,2,*,3] Stack = [1]

2. Reduces "1" to expression "E" based on rule3. Stack = [E]

3. Shift "+" onto stack from input (in anticipation of rule1). Input = [2,*,3] Stack = [E,+]

4. Shift "2" onto stack from input (in anticipation of rule3). Input = [*,3] Stack = [E,+,2]

5. Reduce stack element "2" to Expression "E" based on rule3. Stack = [E,+,E]

6. Reduce stack items [E,+] and new input "E" to "E" based on rule1. Stack = [E]

7. Shift "*" onto stack from input (in anticipation of rule2). Input = Stack = [E,*]

8. Shift "3" onto stack from input (in anticipation of rule3). Input = [] (empty) Stack = [E,*,3]

9. Reduce stack element "3" to expression "E" based on rule3. Stack = [E,*,E]

10. Reduce stack items [E,*] and new input "E" to "E" based on rule2. Stack = [E]

The parse tree and resulting code from it is not correct according to language semantics.

To correctly parse without lookahead, there are three solutions:

- The user has to enclose expressions within parentheses. This often is not a viable solution.

- The parser needs to have more logic to backtrack and retry whenever a rule is violated or not complete. The similar method is followed in LL parsers.

- Alternatively, the parser or grammar needs to have extra logic to delay reduction and reduce only when it is absolutely sure which rule to reduce first. This method is used in LR parsers. This correctly parses the expression but with many more states and increased stack depth.

Lookahead parser actions

1. Shift 1 onto stack on input 1 in anticipation of rule3. It does not reduce immediately.

2. Reduce stack item 1 to simple Expression on input + based on rule3. The lookahead is +, so we are on path to E +, so we can reduce the stack to E.

3. Shift + onto stack on input + in anticipation of rule1.

4. Shift 2 onto stack on input 2 in anticipation of rule3.

5. Reduce stack item 2 to Expression on input * based on rule3. The lookahead * expects only

E before it.

6. Now stack has E + E and still the input is *. It has two choices now, either to shift based on rule2 or reduction based on rule1. Since * has higher precedence than + based on rule4, we shift * onto stack in anticipation of rule2.

7. Shift 3 onto stack on input 3 in anticipation of rule3.

8. Reduce stack item 3 to Expression after seeing end of input based on rule3.

9. Reduce stack items E * E to E based on rule2.

10. Reduce stack items E + E to E based on rule1.

The parse tree generated is correct and simply more efficient than non-lookahead parsers. This is the strategy followed in LALR parsers.

Natural Language Generation

Natural language generation (NLG) is the natural language processing task of generating natural language from a machine representation system such as a knowledge base or a logical form. Psycholinguists prefer the term language production when such formal representations are interpreted as models for mental representations.

It could be said an NLG system is like a translator that converts data into a natural language representation. However, the methods to produce the final language are different from those of a compiler due to the inherent expressivity of natural languages. NLG has existed for a long time but commercial NLG technology has only recently become widely available.

NLG may be viewed as the opposite of natural language understanding: whereas in natural language understanding the system needs to disambiguate the input sentence to produce the machine representation language, in NLG the system needs to make decisions about how to put a concept into words.

A simple example is systems that generate form letters. These do not typically involve grammar rules, but may generate a letter to a consumer, e.g. stating that a credit card spending limit was reached. To put it another way, simple systems use a template not unlike a Word document mail merge, but more complex NLG systems dynamically create text. As in other areas of natural language processing, this can be done using either explicit models of language (e.g., grammars) and the domain, or using statistical models derived by analysing human-written texts.

Example

The *Pollen Forecast for Scotland* system is a simple example of a simple NLG system that could essentially be a template. This system takes as input six numbers, which give predicted pollen levels in different parts of Scotland. From these numbers, the system generates a short textual summary of pollen levels as its output.

For example, using the historical data for 1-July-2005, the software produces

Grass pollen levels for Friday have increased from the moderate to high levels of yesterday with values of around 6 to 7 across most parts of the country. However, in Northern areas, pollen levels will be moderate with values of 4.

In contrast, the actual forecast (written by a human meteorologist) from this data was

Pollen counts are expected to remain high at level 6 over most of Scotland, and even level 7 in the south east. The only relief is in the Northern Isles and far northeast of mainland Scotland with medium levels of pollen count.

Comparing these two illustrates some of the choices that NLG systems must make; these are further discussed below.

Stages

The process to generate text can be as simple as keeping a list of canned text that is copied and pasted, possibly linked with some glue text. The results may be satisfactory in simple domains such as horoscope machines or generators of personalised business letters. However, a sophisticated NLG system needs to include stages of planning and merging of information to enable the generation of text that looks natural and does not become repetitive. The typical stages of natural language generation, as proposed by Dale and Reiter, are:

Content determination: Deciding what information to mention in the text. For instance, in the pollen example above, deciding whether to explicitly mention that pollen level is 7 in the south east.

Document structuring: Overall organisation of the information to convey. For example, deciding to describe the areas with high pollen levels first, instead of the areas with low pollen levels.

Aggregation: Merging of similar sentences to improve readability and naturalness. For instance, merging the two sentences *Grass pollen levels for Friday have increased from the moderate to high levels of yesterday* and *Grass pollen levels will be around 6 to 7 across most parts of the country* into the single sentence *Grass pollen levels for Friday have increased from the moderate to high levels of yesterday with values of around 6 to 7 across most parts of the country.*

Lexical choice: Putting words to the concepts. For example, deciding whether *medium* or *moderate* should be used when describing a pollen level of 4.

Referring expression generation: Creating referring expressions that identify objects and regions. For example, deciding to use *in the Northern Isles and far northeast of mainland Scotland* to refer to a certain region in Scotland. This task also includes making decisions about pronouns and other types of anaphora.

Realisation: Creating the actual text, which should be correct according to the rules of syntax, morphology, and orthography. For example, using *will be* for the future tense of *to be*.

Applications

The popular media has paid the most attention to NLG systems which generate jokes, but from a commercial perspective, the most successful NLG applications have been *data-to-text* systems which generate textual summaries of databases and data sets; these systems usually perform data analysis as well as text generation. In particular, several systems have been built that produce textual weather forecasts from weather data. The earliest such system to be deployed was FoG, which was used by Environment Canada to generate weather forecasts in French and English in the early 1990s. The success of FoG triggered other work, both research and commercial. Recent research in this area include an experiment which showed that users sometimes preferred computer-generated weather forecasts to human-written ones, in part because the computer forecasts used more consistent terminology, and a demonstration that statistical techniques could be used to generate high-quality weather forecasts. Recent applications include the UK Met Office's text-enhanced forecast.

The use of NLG in financial services is growing as well. In 2016, FactSet discussed with Forbes how they use NLG to automatically write thousands of reports.The use case in financial services seems to be growing solutions like Yseop Compose offer use cases on their website for finance.

This all began in the 1990s when there was interest in using NLG to summarise financial and business data. For example, the SPOTLIGHT system developed at A.C. Nielsen automatically generated readable English text based on the analysis of large amounts of retail sales data. More recently there is interest in using NLG to summarise electronic medical records. Commercial applications in this area are appearing, and researchers have shown that NLG summaries of medical data can be effective decision-support aids for medical professionals. There is also growing interest in using NLG to enhance accessibility, for example by describing graphs and data sets to blind people.

An example of an interactive use of NLG is the WYSIWYM framework. It stands for *What you see is what you meant* and allows users to see and manipulate the continuously rendered view (NLG output) of an underlying formal language document (NLG input), thereby editing the formal language without learning it.

Content generation systems assist human writers and makes writing process more efficient and effective. A content generation tool based on web mining using search engines APIs has been built. The tool imitates the cut-and-paste writing scenario where a writer forms its content from various search results. Relevance verification is essential to filter out irrelevant search results; it is based on matching the parse tree of a query with the parse trees of candidate answers. In an alternative approach, a high-level structure of human-authored text is used to automatically build a template for a new topic for automatically written Wikipedia article.

Several companies have been started since 2009 which build systems that transform data into narrative using NLG and AI techniques. These include Arria NLG, Automated Insights, Narrative Science, Narrativa and Yseop.

Evaluation

As in other scientific fields, NLG researchers need to test how well their systems, modules, and algorithms work. This is called *evaluation*. There are three basic techniques for evaluating NLG systems:

- *Task-based (extrinsic) evaluation*: give the generated text to a person, and assess how well it helps him perform a task (or otherwise achieves its communicative goal). For example, a system which generates summaries of medical data can be evaluated by giving these summaries to doctors, and assessing whether the summaries helps doctors make better decisions.

- *Human ratings*: give the generated text to a person, and ask him or her to rate the quality and usefulness of the text.

- *Metrics*: compare generated texts to texts written by people from the same input data, using an automatic metric such as BLEU.

An ultimate goal is how useful NLG systems are at helping people, which is the first of the above techniques. However, task-based evaluations are time-consuming and expensive, and can be difficult to carry out (especially if they require subjects with specialised expertise, such as doctors). Hence (as in other areas of NLP) task-based evaluations are the exception, not the norm.

Recently researchers are assessing how well human-ratings and metrics correlate with (predict) task-based evaluations. Work is being conducted in the context of Generation Challenges shared-task events. Initial results suggest that human ratings are much better than metrics in this regard. In other words, human ratings usually do predict task-effectiveness at least to some degree (although there are exceptions), while ratings produced by metrics often do not predict task-effectiveness well. These results are preliminary. In any case, human ratings are the most popular evaluation technique in NLG; this is contrast to machine translation, where metrics are widely used.

Natural Language Generation

The steps in natural language generation are as follows.

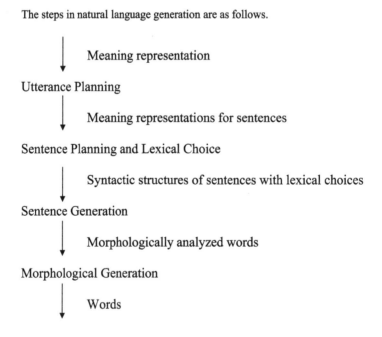

Steps in Language Understanding and Generation

Morphological Analysis

- Analyzing words into their linguistic components (morphemes).

- Morphemes are the smallest meaningful units of language.

cars	car+PLU	
giving	give+PROG	
geliyordum	gel+PROG+PAST+1SG	- I was coming

- Ambiguity: More than one alternatives

flies	flyVERB+PROG	
fly	NOUN+PLU	
adam	adam+ACC	- the man (accusative)
	adam +P1SG	- my man
	ada+P1SG+ACC - my island (accusative)	

Parts-of-Speech (POS) Tagging

- Each word has a part-of-speech tag to describe its category.

- Part-of-speech tag of a word is one of major word groups (or its subgroups).

  - open classes -- noun, verb, adjective, adverb

  - closed classes -- prepositions, determiners, conjuctions, pronouns, particples

- POS Taggers try to find POS tags for the words.

- duck is a verb or noun? (morphological analyzer cannot make decision).

- A POS tagger may make that decision by looking the surrounding words.

  - Duck! (verb)

  - Duck is delicious for dinner. (noun)

Lexical Processing

- The purpose of lexical processing is to determine meanings of individual words.

- Basic methods is to lookup in a database of meanings – lexicon

- We should also identify non-words such as punctuation marks.

- Word-level ambiguity -- words may have several meanings, and the correct one cannot be chosen based solely on the word itself.

  - bank in English

- Solution -- resolve the ambiguity on the spot by POS tagging (if possible) or pass-on the ambiguity to the other levels.

Syntactic Processing

- Parsing -- converting a flat input sentence into a hierarchical structure that corresponds to the units of meaning in the sentence.

- There are different parsing formalisms and algorithms.

- Most formalisms have two main components:

    - grammar -- a declarative representation describing the syntactic structure of sentences in the language.

    - parser -- an algorithm that analyzes the input and outputs its structural representation (its parse) consistent with the grammar specification.

- CFGs are in the center of many of the parsing mechanisms. But they are complemented by some additional features that make the formalism more suitable to handle natural languages.

Semantic Analysis

- Assigning meanings to the structures created by syntactic analysis.

- Mapping words and structures to particular domain objects in way consistent with our knowledge of the world.

- Semantic can play an import role in selecting among competing syntactic analyses and discarding illogical analyses.

    - I robbed the bank -- bank is a river bank or a financial institution

- We have to decide the formalisms which will be used in the meaning representation.

Knowledge Representation for NLP

- Which knowledge representation will be used depends on the application -- Machine Translation, Database Query System.

- Requires the choice of representational framework, as well as the specific meaning vocabulary (what are concepts and relationship between these concepts -- ontology)

- Must be computationally effective.

- Common representational formalisms:

    - first order predicate logic

    - conceptual dependency graphs

    - semantic networks

    - Frame-based representations

Discourse

- Discourses are collection of coherent sentences (not arbitrary set of sentences)

- Discourses have also hierarchical structures (similar to sentences)

- anaphora resolution -- to resolve referring expression

    - Mary bought a book for Kelly. She didn't like it.

        - She refers to Mary or Kelly. -- possibly Kelly

        - It refers to what -- book.

    - Mary had to lie for Kelly. She didn't like it.

- Discourse structure may depend on application.

    - Monologue

    - Dialogue

    - Human-Computer Interaction

Applications of Natural Language Processing

- Machine Translation – Translation between two natural languages.

    - The Babel Fish translations system on Alta Vista.

- Information Retrieval – Web search (uni-lingual or multi-lingual).

- Query Answering/Dialogue – Natural language interface with a database system, or a dialogue system.

- Report Generation – Generation of reports such as weather reports.

- Some Small Applications –

    - Grammar Checking, Spell Checking, Spell Corrector

Machine Translation

Machine translation, sometimes referred to by the abbreviation MT, is a sub-field of computational linguistics that investigates the use of software to translate text or speech from one language to another.

On a basic level, MT performs simple substitution of words in one language for words in another, but that alone usually cannot produce a good translation of a text because recognition of whole phrases and their closest counterparts in the target language is needed. Solving this problem with corpus statistical, and neural techniques is a rapidly growing field that is leading to better translations, handling differences in linguistic typology, translation of idioms, and the isolation of anomalies.

Current machine translation software often allows for customization by domain or profession (such as weather reports), improving output by limiting the scope of allowable substitutions. This technique is particularly effective in domains where formal or formulaic language is used. It follows that machine translation of government and legal documents more readily produces usable output than conversation or less standardised text.

Improved output quality can also be achieved by human intervention: for example, some systems are able to translate more accurately if the user has unambiguously identified which words in the text are proper names. With the assistance of these techniques, MT has proven useful as a tool to assist human translators and, in a very limited number of cases, can even produce output that can be used as is (e.g., weather reports).

The progress and potential of machine translation have been debated much through its history. Since the 1950s, a number of scholars have questioned the possibility of achieving fully automatic machine translation of high quality. Some critics claim that there are in-principle obstacles to automating the translation process.

History

The idea of machine translation may be traced back to the 17th century. In 1629, René Descartes proposed a universal language, with equivalent ideas in different tongues sharing one symbol. The field of "machine translation" appeared in Warren Weaver's Memorandum on Translation (1949). The first researcher in the field, Yehosha Bar-Hillel, began his research at MIT (1951). A Georgetown University MT research team followed (1951) with a public demonstration of its Georgetown-IBM experiment system in 1954. MT research programs popped up in Japan and Russia (1955), and the first MT conference was held in London (1956). Researchers continued to join the field as the Association for Machine Translation and Computational Linguistics was formed in the U.S. (1962) and the National Academy of Sciences formed the Automatic Language Processing Advisory Committee (ALPAC) to study MT (1964). Real progress was much slower, however, and after the ALPAC report (1966), which found that the ten-year-long research had failed to fulfill expectations, funding was greatly reduced. According to a 1972 report by the Director of Defense Research and Engineering (DDR&E), the feasibility of large-scale MT was reestablished by the success of the Logos MT system in translating military manuals into Vietnamese during that conflict.

The French Textile Institute also used MT to translate abstracts from and into French, English, German and Spanish (1970); Brigham Young University started a project to translate Mormon texts by automated translation (1971); and Xerox used SYSTRAN to translate technical manuals (1978). Beginning in the late 1980s, as computational power increased and became less expensive, more interest was shown in statistical models for machine translation. Various MT companies were launched, including Trados (1984), which was the first to develop and market translation memory technology (1989). The first commercial MT system for Russian / English / German-Ukrainian was developed at Kharkov State University (1991).

MT on the web started with SYSTRAN Offering free translation of small texts (1996), followed by AltaVista Babelfish, which racked up 500,000 requests a day (1997). Franz-Josef Och (the future head of Translation Development AT Google) won DARPA's speed MT competition (2003). More innovations during this time included MOSES, the open-source statistical MT engine (2007), a text/SMS translation service for mobiles in Japan (2008), and a mobile phone with built-in speech-to-speech translation functionality for English, Japanese and Chinese (2009). Recently, Google announced that Google Translate translates roughly enough text to fill 1 million books in one day (2012).

The idea of using digital computers for translation of natural languages was proposed as early as 1946 by A. D. Booth and possibly others. Warren Weaver wrote an important memorandum "Translation" in 1949. The Georgetown experiment was by no means the first such application, and a demonstration was made in 1954 on the APEXC machine at Birkbeck College (University of London) of a rudimentary translation of English into French. Several papers on the topic were published at the time, and even articles in popular journals (for example *Wireless World*, Sept. 1955, Cleave and Zacharov). A similar application, also pioneered at Birkbeck College at the time, was reading and composing Braille texts by computer.

Translation Process

The human translation process may be described as:

1. Decoding the meaning of the source text; and

2. Re-encoding this meaning in the target language.

Behind this ostensibly simple procedure lies a complex cognitive operation. To decode the meaning of the source text in its entirety, the translator must interpret and analyse all the features of the text, a process that requires in-depth knowledge of the grammar, semantics, syntax, idioms, etc., of the source language, as well as the culture of its speakers. The translator needs the same in-depth knowledge to re-encode the meaning in the target language.

Therein lies the challenge in machine translation: how to program a computer that will "understand" a text as a person does, and that will "create" a new text in the target language that "sounds" as if it has been written by a person.

In its most general application, this is beyond current technology. Though it works much faster, no automated translation program or procedure, with no human participation, can produce output even close to the quality a human translator can produce. What it can do, however, is provide a general, though imperfect, approximation of the original text, getting the "gist" of it (a process called "gisting"). This is sufficient for many purposes, including making best use of the finite and expensive time of a human translator, reserved for those cases in which total accuracy is indispensable.

This problem may be approached in a number of ways, through the evolution of which accuracy has improved.

Approaches

Machine translation can use a method based on linguistic rules, which means that words will be translated in a linguistic way – the most suitable (orally speaking) words of the target language will replace the ones in the source language.

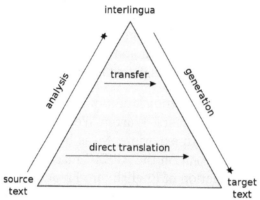

Bernard Vauquois' pyramid showing comparative depths of intermediary representation, interlingual machine translation at the peak, followed by transfer-based, then direct translation.

It is often argued that the success of machine translation requires the problem of natural language understanding to be solved first.

Generally, rule-based methods parse a text, usually creating an intermediary, symbolic representation, from which the text in the target language is generated. According to the nature of the intermediary representation, an approach is described as interlingual machine translation or transfer-based machine translation. These methods require extensive lexicons with morphological, syntactic, and semantic information, and large sets of rules.

Given enough data, machine translation programs often work well enough for a native speaker of one language to get the approximate meaning of what is written by the other native speaker. The difficulty is getting enough data of the right kind to support the particular method. For example, the large multilingual corpus of data needed for statistical methods to work is not necessary for the grammar-based methods. But then, the grammar methods need a skilled linguist to carefully design the grammar that they use.

To translate between closely related languages, the technique referred to as rule-based machine translation may be used.

Rule-based

The rule-based machine translation paradigm includes transfer-based machine translation, interlingual machine translation and dictionary-based machine translation paradigms. This type of translation is used mostly in the creation of dictionaries and grammar programs. Unlike other methods, RBMT involves more information about the linguistics of the source and target languages, using the morphological and syntactic rules and semantic analysis of both languages. The basic approach involves linking the structure of the input sentence with the structure of the output sentence using a parser and an analyzer for the source language, a generator for the target language, and a transfer lexicon for the actual translation. RBMT's biggest downfall is that everything must be made explicit: orthographical variation and erroneous input must be made part of the source language analyser in order to cope with it, and lexical selection rules must be written for all instances of ambiguity. Adapting to new domains in itself is not that hard, as the core grammar is the same across domains, and the domain-specific adjustment is limited to lexical selection adjustment.

Transfer-based Machine Translation

Transfer-based machine translation is similar to interlingual machine translation in that it creates a translation from an intermediate representation that simulates the meaning of the original sentence. Unlike interlingual MT, it depends partially on the language pair involved in the translation.

Interlingual

Interlingual machine translation is one instance of rule-based machine-translation approaches. In this approach, the source language, i.e. the text to be translated, is transformed into an interlingual language, i.e. a "language neutral" representation that is independent of any language. The target language is then generated out of the interlingua. One of the major advantages of this system is that the interlingua becomes more valuable as the number of target languages it can be turned into increases. However, the only interlingual machine translation system that has been made operational at the commercial level is the KANT system (Nyberg and Mitamura, 1992), which is designed to translate Caterpillar Technical English (CTE) into other languages.

Dictionary-based

Machine translation can use a method based on dictionary entries, which means that the words will be translated as they are by a dictionary.

Statistical

Statistical machine translation tries to generate translations using statistical methods based on bilingual text corpora, such as the Canadian Hansard corpus, the English-French record of the Canadian parliament and EUROPARL, the record of the European Parliament. Where such corpora are available, good results can be achieved translating similar texts, but such corpora are still rare for many language pairs. The first statistical machine translation software was CANDIDE from IBM. Google used SYSTRAN for several years, but switched to a statistical translation method in October 2007. In 2005, Google improved its internal translation capabilities by using approximately 200 billion words from United Nations materials to train their system; translation accuracy improved. Google Translate and similar statistical translation programs work by detecting patterns in hundreds of millions of documents that have previously been translated by humans and making intelligent guesses based on the findings. Generally, the more human-translated documents available in a given language, the more likely it is that the translation will be of good quality. Newer approaches into Statistical Machine translation such as METIS II and PRESEMT use minimal corpus size and instead focus on derivation of syntactic structure through pattern recognition. With further development, this may allow statistical machine translation to operate off of a monolingual text corpus. SMT's biggest downfall includes it being dependent upon huge amounts of parallel texts, its problems with morphology-rich languages (especially with translating *into* such languages), and its inability to correct singleton errors.

Example-based

Example-based machine translation (EBMT) approach was proposed by Makoto Nagao in 1984. Example-based machine translation is based on the idea of analogy. In this approach, the corpus that is used is one that contains texts that have already been translated. Given a sentence that is to be translated, sentences from this corpus are selected that contain similar sub-sentential components. The similar sentences are then used to translate the sub-sentential components of the original sentence into the target language, and these phrases are put together to form a complete translation.

Hybrid MT

Hybrid machine translation (HMT) leverages the strengths of statistical and rule-based translation methodologies. Several MT organizations (such as Omniscien Technologies (formerly Asia Online), LinguaSys, Systran, and Polytechnic University of Valencia) claim a hybrid approach that uses both rules and statistics. The approaches differ in a number of ways:

- Rules post-processed by statistics: Translations are performed using a rules based engine. Statistics are then used in an attempt to adjust/correct the output from the rules engine.

- Statistics guided by rules: Rules are used to pre-process data in an attempt to better guide the statistical engine. Rules are also used to post-process the statistical output to perform functions such as normalization. This approach has a lot more power, flexibility and control when translating.

Neural MT

A deep learning based approach to MT, neural machine translation has made rapid progress in recent years, and Google has announced its translation services are now using this technology in preference to its previous statistical methods.

Major Issues

Disambiguation

Word-sense disambiguation concerns finding a suitable translation when a word can have more than one meaning. The problem was first raised in the 1950s by Yehoshua Bar-Hillel. He pointed out that without a "universal encyclopedia", a machine would never be able to distinguish between the two meanings of a word. Today there are numerous approaches designed to overcome this problem. They can be approximately divided into "shallow" approaches and "deep" approaches.

Shallow approaches assume no knowledge of the text. They simply apply statistical methods to the words surrounding the ambiguous word. Deep approaches presume a comprehensive knowledge of the word. So far, shallow approaches have been more successful.

Claude Piron, a long-time translator for the United Nations and the World Health Organization, wrote that machine translation, at its best, automates the easier part of a translator's job; the harder and more time-consuming part usually involves doing extensive research to resolve ambiguities in the source text, which the grammatical and lexical exigencies of the target language require to be resolved:

> Why does a translator need a whole workday to translate five pages, and not an hour or two? ..... About 90% of an average text corresponds to these simple conditions. But unfortunately, there's the other 10%. It's that part that requires six [more] hours of work. There are ambiguities one has to resolve. For instance, the author of the source text, an Australian physician, cited the example of an epidemic which was declared during World War II in a "Japanese prisoner of war camp". Was he talking about an American camp with Japanese prisoners or a Japanese camp with American prisoners? The English has two senses. It's necessary therefore to do research, maybe to the extent of a phone call to Australia.

The ideal deep approach would require the translation software to do all the research necessary for this kind of disambiguation on its own; but this would require a higher degree of AI than has yet been attained. A shallow approach which simply guessed at the sense of the ambiguous English phrase that Piron mentions (based, perhaps, on which kind of prisoner-of-war camp is more often mentioned in a given corpus) would have a reasonable chance of guessing wrong fairly often. A shallow approach that involves "ask the user about each ambiguity" would, by Piron's estimate, only automate about 25% of a professional translator's job, leaving the harder 75% still to be done by a human.

Non-standard Speech

One of the major pitfalls of MT is its inability to translate non-standard language with the same accuracy as standard language. Heuristic or statistical based MT takes input from various sources in standard form of a language. Rule-based translation, by nature, does not include common non-standard usages. This causes errors in translation from a vernacular source or into colloquial language. Limitations on translation from casual speech present issues in the use of machine translation in mobile devices.

Named Entities

Name entities, in narrow sense, refer to concrete or abstract entities in the real world including people, organizations, companies, places etc. It also refers to expressing of time, space, quantity such as 1 July 2011, $79.99 and so on.

Named entities occur in the text being analyzed in statistical machine translation. The initial difficulty that arises in dealing with named entities is simply identifying them in the text. Consider the list of names common in a particular language to illustrate this – the most common names are different for each language and also are constantly changing. If named entities cannot be recognized by the machine translator, they may be erroneously translated as common nouns, which would most likely not affect the BLEU rating of the translation but would change the text's human readability. It is also possible that, when not identified, named entities will be omitted from the output translation, which would also have implications for the text's readability and message.

Another way to deal with named entities is to use transliteration instead of translation, meaning that you find the letters in the target language that most closely correspond to the name in the source language. There have been attempts to incorporate this into machine translation by adding a transliteration step into the translation procedure. However, these attempts still have their problems and have even been cited as worsening the quality of translation. Named entities were still identified incorrectly, with words not being transliterated when they should or being transliterated when they shouldn't. For example, for "Southern California" the first word should be translated directly, while the second word should be transliterated. However, machines would often transliterate both because they treated them as one entity. Words like these are hard for machine translators, even those with a transliteration component, to process.

The lack of attention to the issue of named entity translation has been recognized as potentially stemming from a lack of resources to devote to the task in addition to the complexity of creating a good system for named entity translation. One approach to named entity translation has been to transliterate, and not translate, those words. A second is to create a "do-not-translate" list, which has the same end goal – transliteration as opposed to translation. Both of these approaches still rely on the correct identification of named entities, however.

A third approach to successful named entity translation is a class-based model. In this method, named entities are replaced with a token to represent the class they belong to. For example, "Ted" and "Erica" would both be replaced with "person" class token. In this way the statistical distribution and use of person names in general can be analyzed instead of looking at the distributions of "Ted" and "Erica" individually. A problem that the class based model solves is that the probability of a given name in a specific language will not affect the assigned probability of a translation. A study by Stanford on improving this area of translation gives the examples that different probabilities will be assigned to "David is going for a walk" and "Ankit is going for a walk" for English as a target language due to the different number of occurrences for each name in the training data. A frustrating outcome of the same study by Stanford (and other attempts to improve named recognition translation) is that many times, a decrease in the BLEU scores for translation will result from the inclusion of methods for named entity translation.

Translation from Multiparallel Sources

Some work has been done in the utilization of multiparallel corpora, that is a body of text that has been translated into 3 or more languages. Using these methods, a text that has been translated into 2 or more languages may be utilized in combination to provide a more accurate translation into a third language compared with if just one of those source languages were used alone.

Ontologies in MT

An ontology is a formal representation of knowledge which includes the concepts (such as objects, processes etc.) in a domain and some relations between them. If the stored information is of linguistic nature, one can speak of a lexicon. In NLP, ontologies can be used as a source of knowledge for machine translation systems. With access to a large knowledge base, systems can be enabled to resolve many (especially lexical) ambiguities on their own. In the following classic examples, as humans, we are able to interpret the prepositional phrase according to the context because we use our world knowledge, stored in our lexicons:

"I saw a man/star/molecule with a microscope/telescope/binoculars."

A machine translation system initially would not be able to differentiate between the meanings because syntax does not change. With a large enough ontology as a source of knowledge however, the possible interpretations of ambiguous words in a specific context can be reduced. Other areas of usage for ontologies within NLP include information retrieval, information extraction and text summarization.

Building Ontologies

The ontology generated for the PANGLOSS knowledge-based machine translation system in 1993 may serve as an example of how an ontology for NLP purposes can be compiled:

- A large-scale ontology is necessary to help parsing in the active modules of the machine translation system.

- In the PANGLOSS example, about 50.000 nodes were intended to be subsumed under the smaller, manually-built *upper* (abstract) *region* of the ontology. Because of its size, it had to be created automatically.

- The goal was to merge the two resources LDOCE online and WordNet to combine the benefits of both: concise definitions from Longman, and semantic relations allowing for semi-automatic taxonomization to the ontology from WordNet.

  o A *definition match* algorithm was created to automatically merge the correct meanings of ambiguous words between the two online resources, based on the words that the definitions of those meanings have in common in LDOCE and WordNet. Using a similarity matrix, the algorithm delivered matches between meanings including a confidence factor. This algorithm alone, however, did not match all meanings correctly on its own.

  o A second *hierarchy match* algorithm was therefore created which uses the taxonomic hierarchies found in WordNet (deep hierarchies) and partially in LDOCE (flat hierarchies). This works by first matching unambiguous meanings, then limiting the search space to only the respective ancestors and descendants of those matched meanings. Thus, the algorithm matched locally unambiguous meanings (for instance, while the word *seal* as such is ambiguous, there is only one meaning of "seal" in the *animal* subhierarchy).

- Both algorithms complemented each other and helped constructing a large-scale ontology for the machine translation system. The WordNet hierarchies, coupled with the matching definitions of LDOCE, were subordinated to the ontology's *upper region*. As a result, the PANGLOSS MT system was able to make use of this knowledge base, mainly in its generation element.

Applications

While no system provides the holy grail of fully automatic high-quality machine translation of unrestricted text, many fully automated systems produce reasonable output. The quality of machine translation is substantially improved if the domain is restricted and controlled.

Despite their inherent limitations, MT programs are used around the world. Probably the largest institutional user is the European Commission. The MOLTO project, for example, coordinated by the University of Gothenburg, received more than 2.375 million euros project support from the EU to create a reliable translation tool that covers a majority of the EU languages. The further development of MT systems comes at a time when budget cuts in human translation may increase the EU's dependency on reliable MT programs. The European Commission contributed 3.072 million euros (via its ISA programme) for the creation of MT@EC, a statistical machine translation program tailored to the administrative needs of the EU, to replace a previous rule-based machine translation system.

Google has claimed that promising results were obtained using a proprietary statistical machine translation engine. The statistical translation engine used in the Google language tools for Arabic <-> English and Chinese <-> English had an overall score of 0.4281 over the runner-up IBM's BLEU-4 score of 0.3954 (Summer 2006) in tests conducted by the National Institute for Standards and Technology.

With the recent focus on terrorism, the military sources in the United States have been investing significant amounts of money in natural language engineering. *In-Q-Tel* (a venture capital fund, largely funded by the US Intelligence Community, to stimulate new technologies through private sector entrepreneurs) brought up companies like Language Weaver. Currently the military community is interested in translation and processing of languages like Arabic, Pashto, and Dari. Within these languages, the focus is on key phrases and quick communication between military members and civilians through the use of mobile phone apps. The Information Processing Technology Office in DARPA hosts programs like TIDES and Babylon translator. US Air Force has awarded a $1 million contract to develop a language translation technology.

The notable rise of social networking on the web in recent years has created yet another niche for the application of machine translation software – in utilities such as Facebook, or instant messaging clients such as Skype, GoogleTalk, MSN Messenger, etc. – allowing users speaking different languages to communicate with each other. Machine translation applications have also been released for most mobile devices, including mobile telephones, pocket PCs, PDAs, etc. Due to their portability, such instruments have come to be designated as mobile translation tools enabling mobile business networking between partners speaking different languages, or facilitating both foreign language learning and unaccompanied traveling to foreign countries without the need of the intermediation of a human translator.

Despite being labelled as an unworthy competitor to human translation in 1966 by the Automated Language Processing Advisory Committee put together by the United States government, the quality of machine translation has now been improved to such levels that its application in online collaboration and in the medical field are being investigated. In the Ishida and Matsubara lab of Kyoto University, methods of improving the accuracy of machine translation as a support tool for inter-cultural collaboration in today's globalized society are being studied. The application of this technology in medical settings where human translators are absent is another topic of research however difficulties arise due to the importance of accurate translations in medical diagnoses.

Evaluation

There are many factors that affect how machine translation systems are evaluated. These factors include the intended use of the translation, the nature of the machine translation software, and the nature of the translation process.

Different programs may work well for different purposes. For example, statistical machine translation (SMT) typically outperforms example-based machine translation (EBMT), but researchers found that when evaluating English to French translation, EBMT performs better. The same concept applies for technical documents, which can be more easily translated by SMT because of their formal language.

In certain applications, however, e.g., product descriptions written in a controlled language, a dictionary-based machine-translation system has produced satisfactory translations that require no human intervention save for quality inspection.

There are various means for evaluating the output quality of machine translation systems. The oldest is the use of human judges to assess a translation's quality. Even though human evaluation is time-consuming, it is still the most reliable method to compare different systems such as rule-based and statistical systems. Automated means of evaluation include BLEU, NIST, METEOR, and LEPOR.

Relying exclusively on unedited machine translation ignores the fact that communication in human language is context-embedded and that it takes a person to comprehend the context of the original text with a reasonable degree of probability. It is certainly true that even purely human-generated translations are prone to error. Therefore, to ensure that a machine-generated translation will be useful to a human being and that publishable-quality translation is achieved, such translations must be reviewed and edited by a human. The late Claude Piron wrote that machine translation, at its best, automates the easier part of a translator's job; the harder and more time-consuming part usually involves doing extensive research to resolve ambiguities in the source text, which the grammatical and lexical exigencies of the target language require to be resolved. Such research is a necessary prelude to the pre-editing necessary in order to provide input for machine-translation software such that the output will not be meaningless.

In addition to disambiguation problems, decreased accuracy can occur due to varying levels of training data for machine translating programs. Both example-based and statistical machine translation rely on a vast array of real example sentences as a base for translation, and when too many or too few sentences are analyzed accuracy is jeopardized. Researchers found that when a program is trained on 203,529 sentence pairings, accuracy actually decreases. The optimal level of training data seems to be just over 100,000 sentences, possibly because as training data increasing, the number of possible sentences increases, making it harder to find an exact translation match.

Using Machine Translation as a Teaching Tool

Although there have been concerns about machine translation's accuracy, Dr. Ana Nino of the University of Manchester has researched some of the advantages in utilizing machine translation in the classroom. One such pedagogical method is called using "MT as a Bad Model." MT as a Bad Model forces the language learner to identify inconsistencies or incorrect aspects of a translation; in turn, the individual will (hopefully) possess a better grasp of the language. Dr. Nino cites that this teaching tool was implemented in the late 1980s. At the end of various semesters, Dr. Nino was able to obtain survey results from students who had used MT as a Bad Model (as well as other models.) Overwhelmingly, students felt that they had observed improved comprehension, lexical retrieval, and increased confidence in their target language.

Machine Translation and Signed Languages

In the early 2000s, options for machine translation between spoken and signed languages were severely limited. It was a common belief that deaf individuals could use traditional translators. However, stress, intonation, pitch, and timing are conveyed much differently in spoken languages compared to signed languages. Therefore, a deaf individual may misinterpret or become confused about the meaning of written text that is based on a spoken language.

Researchers Zhao, et al. (2000), developed a prototype called TEAM (translation from English to ASL by machine) that completed English to American Sign Language (ASL) translations. The program would first analyze the syntactic, grammatical, and morphological aspects of the English text. Following this step, the program accessed a sign synthesizer, which acted as a dictionary for ASL. This synthesizer housed the process one must follow to complete ASL signs, as well as the meanings of these signs. Once the entire text is analyzed and the signs necessary to complete the translation are located in the synthesizer, a computer generated human appeared and would use ASL to sign the English text to the user.

Copyright

Only works that are original are subject to copyright protection, so some scholars claim that machine translation results are not entitled to copyright protection because MT does not involve creativity. The copyright at issue is for a derivative work; the author of the original work in the original language does not lose his rights when a work is translated: a translator must have permission to publish a translation.

References

- Galitsky, Boris (2013). "A Web Mining Tool for Assistance with Creative Writing". Advances in Information Retrieval. Lecture Notes in Computer Science. 7814: 828–831. doi:10.1007/978-3-642-36973-5_95

- Alisa Kongthon, Chatchawal Sangkeettrakarn, Sarawoot Kongyoung and Choochart Haruechaiyasak. Published by ACM 2009 Article, Bibliometrics Data Bibliometrics. Published in: Proceeding, MEDES '09 Proceedings of the International Conference on Management of Emergent Digital EcoSystems, ACM New York, NY, USA. ISBN 978-1-60558-829-2

- Law A, Freer Y, Hunter J, Logie R, McIntosh N, Quinn J (2005). "A Comparison of Graphical and Textual Presentations of Time Series Data to Support Medical Decision Making in the Neonatal Intensive Care Unit". Journal of Clinical Monitoring and Computing. 19 (3): 183–94

- Dale, Robert; Reiter, Ehud (2000). Building natural language generation systems. Cambridge, U.K.: Cambridge University Press. ISBN 0-521-02451-X

- Goldberg, Yoav (2016) A Primer on Neural Network Models for Natural Language Processing. Journal of Artificial Intelligence Research 57 (2016) 345–420

- Anand, Tej; Kahn, Gary (1992). "Making Sense of Gigabytes: A System for Knowledge-Based Market Analysis" (PDF). In Klahr, Philip; Scott, A. F. Innovative applications of artificial intelligence 4: proceedings of the IAAI-92 Conference. Menlo Park, Calif: AAAI Press. pp. 57–70. ISBN 0-262-69155-8

- Yucong Duan, Christophe Cruz (2011), Formalizing Semantic of Natural Language through Conceptualization from Existence. International Journal of Innovation, Management and Technology(2011) 2 (1), pp. 37-42

Permissions

All chapters in this book are published with permission under the Creative Commons Attribution Share Alike License or equivalent. Every chapter published in this book has been scrutinized by our experts. Their significance has been extensively debated. The topics covered herein carry significant information for a comprehensive understanding. They may even be implemented as practical applications or may be referred to as a beginning point for further studies.

We would like to thank the editorial team for lending their expertise to make the book truly unique. They have played a crucial role in the development of this book. Without their invaluable contributions this book wouldn't have been possible. They have made vital efforts to compile up to date information on the varied aspects of this subject to make this book a valuable addition to the collection of many professionals and students.

This book was conceptualized with the vision of imparting up-to-date and integrated information in this field. To ensure the same, a matchless editorial board was set up. Every individual on the board went through rigorous rounds of assessment to prove their worth. After which they invested a large part of their time researching and compiling the most relevant data for our readers.

The editorial board has been involved in producing this book since its inception. They have spent rigorous hours researching and exploring the diverse topics which have resulted in the successful publishing of this book. They have passed on their knowledge of decades through this book. To expedite this challenging task, the publisher supported the team at every step. A small team of assistant editors was also appointed to further simplify the editing procedure and attain best results for the readers.

Apart from the editorial board, the designing team has also invested a significant amount of their time in understanding the subject and creating the most relevant covers. They scrutinized every image to scout for the most suitable representation of the subject and create an appropriate cover for the book.

The publishing team has been an ardent support to the editorial, designing and production team. Their endless efforts to recruit the best for this project, has resulted in the accomplishment of this book. They are a veteran in the field of academics and their pool of knowledge is as vast as their experience in printing. Their expertise and guidance has proved useful at every step. Their uncompromising quality standards have made this book an exceptional effort. Their encouragement from time to time has been an inspiration for everyone.

The publisher and the editorial board hope that this book will prove to be a valuable piece of knowledge for students, practitioners and scholars across the globe.

Index